# Top 20 Essential Skills for ArcGIS® Business Analyst™

# TOP 20

## ESSENTIAL SKILLS FOR ARCGIS BUSINESS ANALYST

GEMMA GOODALE-SUSSEN
HELEN THOMPSON

**Esri Press**
Redlands, California

Esri Press, 380 New York Street, Redlands, California 92373-8100

Printed in the United States of America.
30 29 28 27 26    1 2 3 4 5 6 7 8 9 10

Paperback ISBN: 9781589488250
Hardcover ISBN: 9781589489219
E-book ISBN: 9781589488267

Library of Congress Control Number: 2026940637



For purchasing and distribution options (both domestic and international), please visit esripress.esri.com.

# CONTENTS

# PREFACE

Business analytics influences all aspects of modern retail, real estate, financial services, and commercial industries. Every coffee shop, grocery store, bank, and restaurant operate based on market potential assessments, competitor impact analysis, sales forecasts, and customer suitability evaluations. Decisions about opening, expanding, shrinking, or relocating stores rely heavily on location intelligence—insights gained by visualizing, integrating, and analyzing geospatial data and software with an organization's information and market expertise. For more than 40 years, location intelligence has helped clarify what is happening in a particular area, why it is occurring, and how it impacts local businesses and communities.

Today, location intelligence also underpins decisions in manufacturing and supply chains, health care, hospitality, construction, entertainment, and more. It's an essential part of every Fortune 1,000 business and yet accessible enough to support an independent franchisee—say, with only one location in suburban Atlanta, for example—in growing their business. It enables local governments to invest in economic development, supports public safety organizations to keep our communities safe, and ensures that products and services are appropriately aligned with the needs of citizens and consumers.

Location intelligence grew out of more than a century of work in economic geography, which aimed to quantify and understand the spatial patterns of production, distribution, and customer interactions. It owes its existence to groundbreaking contributions from location theorists, such as Johann Heinrich von Thünen, Walter Christaller, William Reilly, Waldo Tobler, David Huff, Micheal Batty, and others. Now, in 2026, location intelligence is a widely recognized subset of the geographic information system (GIS) market with dedicated software products, specialized data, and a community of specialized academic and professional researchers.

Business analytics and location intelligence have experienced unprecedented growth over the past four decades. Advances in technology, such as GPS and tracking human movement, now enable us to analyze people's travel patterns with remarkable accuracy in both time and place. Today, it's possible to pinpoint a store and identify exactly who has visited it during a chosen time frame. By examining the demographics of nearby areas where these visitors originate, we can determine how often people visit a location and the total number of visitors from different locations. Comparing these trends provides valuable quantitative insights into visitor behaviors, traits, attitudes, and psychographic profiles.

Segmentation systems enable a more comprehensive understanding of community diversity. Instead of relying solely on metrics, such as median household income or average home price, segmentation facilitates grouping individuals based on shared lifestyles, life stages, and geographic location. Tools such as ArcGIS® Tapestry™ offer a structured approach for characterizing communities, connecting diverse attributes—such as education, housing type, family structure, construction, consumer behavior, attitudes, and aspirations—to identifiable segments. Through analysis of community composition, it is possible to model both commonalities and distinctions across groups, providing insight into how products or services may meet the needs of specific individuals or segment clusters. Segmentation analysis supports strategic targeting by profiling markets, identifying optimal customers, and pinpointing locations where similar high-value customers are concentrated.

Customer relationship management (CRM) systems and point-of-sale (POS) analytics provide valuable insights into store attractiveness and customer draw. Organizations increasingly integrate advanced demographic and geospatial data with their internal platforms, using a comprehensive understanding of their customers to inform market expansion strategies, optimize networks, and configure store layouts.

Over the past decade, location intelligence has broadened its scope to incorporate these technological advancements and beyond. This book brings many of these concepts and advances together to show how ArcGIS Business Analyst™ and its datasets can help businesses and GIS professionals apply location intelligence to business decision-making.

# ACKNOWLEDGMENTS

This book would not have been possible without the significant contribution of others, particularly our colleagues in software development, product engineering, product management, and data production.

Sarah David made an enormous contribution to this book. Her chapters, which focus on Business Analyst Web App, demonstrate her commitment to providing customers with practical knowledge rooted in real-world examples. Sarah's work on blogs and communication strategies has played a pivotal role in broadening the scope and quality of our internal resources, public documentation, and related literature.

Sarah's enthusiasm, dedication, and writing style illuminate both the power and the intricacies of software and data, contributing greatly to both internal and external communications, all while bringing our software and data to life.

Garry Burgess, Max Sattarov, Kyle Watson, Ding Yuan, Cheryl Hagevik, Daniel Stauning, and many others work tirelessly to develop, test, and improve the ArcGIS Business Analyst suite in all its forms. Their dedication to customer service and quality assurance continues to inspire our community to do more.

Christy DeBernardi, Linwei Zeng, Brenda Wolfe, and Harshi Dondapati continue to shape our product strategy and future direction. Collectively, this team has enabled us to transform our organization and create software that is truly driven by user needs.

Stacy Schaetzl, Sandy Silva, Kyle Cassal, Doug Skuta, and their teams continue to be strong partners in data innovation, development, and production, ensuring that ArcGIS Business Analyst data is global in scope and hyperlocal in scale.

Helen: To Clint, for being my mentor, the first renaissance product engineer and friend. To Sara, for the 40-year life journey to three continents and always letting me be me.

Gemma: Many thanks to Molly Zurn for her inspiration and support and to Maryam Mafuri for her immense help with this book.

# INTRODUCTION

## What is ArcGIS Business Analyst?

ArcGIS Business Analyst is powerful location intelligence–based market, customer, and community analytics software. It allows users to combine their own data with thousands of Esri®-curated demographic, business, lifestyle, spending, and census data variables for more than 170 countries worldwide, with map-based analytics for smarter decision-making.

ArcGIS Business Analyst assists users in making more informed data-driven decisions, optimizing operations, and gaining a competitive edge through dedicated workflows, outcome-centric walk-throughs, and geoprocessing toolboxes. As a result, it has been widely adopted by retailers, banks, and manufacturers seeking the best site for a new facility; health-care providers aiming to expand services; government agencies planning economic development; and emergency services planning for, mitigating, and responding to community needs or a disaster.

ArcGIS Business Analyst provides the tools and data needed to turn complex information into actionable insights. *Top 20 Essential Skills for ArcGIS Business Analyst* highlights the role of spatial analysis in making more informed and data-driven decisions. The skills and use cases described in this book are drawn from many real-world scenarios from organizations in retail, manufacturing, real estate, banking, and beyond. We hope that it helps you learn and then apply these skills in your daily work. In our experience, enhancing your demonstrable technical skills and knowledge of spatial analysis offers professional development and career advancement opportunities at a time when companies are increasingly looking for strategic thinkers with business acumen and proven experience.

### Understanding the system: Core concepts

At its heart, ArcGIS Business Analyst uses spatial analysis to help users answer questions related to location: Where are my best customers, and where can I find more of them? What are the demographic characteristics of a neighborhood? How are they changing? Where should I open my next branch to maximize my return on investment (ROI)? What factors are causing the difference in sales between two similar locations? How do these locations compare, and what characteristics are driving customer engagement and sales?

The system combines GIS technology with curated datasets—such as population, income, consumer spending, housing characteristics, lifestyles, market potential, and business locations—allowing users to gain deeper insights into the trends, characteristics, and dynamics of markets, communities, and opportunities. Its intuitive mapping and visualization capabilities, including boardroom-ready infographics, make it accessible to users ranging from GIS professionals to market analysts and executives.

## Software options: Flexible deployment for every size and type of organization

ArcGIS Business Analyst is available in several software formats to suit different organizational needs and technical environments:

- **ArcGIS Business Analyst℠ Web App**: A cloud-based solution, the web app is accessible from any internet-connected device. It offers interactive mapping, analysis, data fusion, and reporting tools without the need for local software installation. Users can quickly generate site reports, perform market analysis, and share results with colleagues. Sharing includes simple exchange formats, such as Excel or PDF; dynamic HTML reports that can be embedded in websites; or sharing through other products, including ArcGIS Dashboards, ArcGIS Storymaps℠, and ArcGIS Experience Builder.
- **ArcGIS Business Analyst™ Pro**: An ArcGIS Pro extension for advanced users and GIS professionals, providing extensive desktop capabilities, including geoprocessing toolboxes, step-by-step wizards, and dedicated workflow support. It allows for location analytics alongside in-depth spatial analysis, custom modeling, and integration with other ArcGIS Pro workflows. Business Analyst Pro provides unique tools for territory design, retail gravity (Huff) modeling, target marketing, and advanced trade area analysis. Desktop users can access locally stored or ArcGIS-hosted data, create tailored analyses, and develop custom tools to automate their modeling processes.
- **ArcGIS Business Analyst™ Mobile App**: Designed for fieldwork and on-the-go analysis, the mobile app enables users to collect site-specific data, view demographic information in real time, and capture photos or notes on phones and tablets. This is ideal for site visits, field surveys, and remote assessments.

It can also be accessed as a web app, providing reporting, site creation, comparison, and analysis for users who prefer a stripped-down experience.

- **ArcGIS Business Analyst™ Enterprise**: For organizations needing enterprise-level deployment, hosted in their organization with scalable, secure access to ArcGIS Business Analyst tools and data across departments. It provides integration with other business systems and custom application development using the underlying ArcGIS Enterprise framework. It provides a dedicated implementation of Business Analyst Web App hosted in the organization and secure access to Business Analyst Pro and Business Analyst Web App for users who need to work with private, confidential, or proprietary information.

With these options, ArcGIS Business Analyst can support a single customer or be tailored for small businesses, multinational corporations, government agencies, and nonprofits alike.

## Types of functionality: From mapping to market analysis

ArcGIS Business Analyst offers a wide array of analytic options and dedicated tools and workflows, uniquely designed for the needs of commercial users and related industries. The software reflects the role of a business analyst rather than being tied to specific industries, such as retail, real estate, or banking. A business analyst uses data analytics to create insights that identify opportunities, issues, and gaps within an organization. Business analysts are found in many organizations and industry sectors. A business analyst's primary focus is to recommend solutions and strategies that improve operational performance and bring about change in how the organization operates.

ArcGIS Business Analyst capabilities are broadly aligned and categorized with the types of analytics and business problem solving required to improve operational performance and maximize effectiveness. ArcGIS Business Analyst functionality includes the following:

- **Site selection and suitability analysis**: Identify optimal spots for new locations, such as stores, clinics, fire stations, or offices, by analyzing demographic trends, competition, traffic patterns, community needs, and market potential.
- **Market segmentation**: Divide populations into distinct groups based on behaviors, preferences, or socioeconomic factors. This enables organizations to improve marketing outreach, targeting, or service delivery.

- **Trade area analysis**: Define and analyze the geographic areas from which customers are drawn. Users can assess drive times, walk times, or custom boundaries to understand market reach, assess competitive threats, or optimize their network of locations to maximize coverage and minimize wasted expenditure.
- **Demographic and consumer insights**: Access rich datasets on population, age, income, education, spending, and lifestyles. These insights inform product launches, expansion plans, store format and differentiation, tenant mix, property reuse, and community initiatives.
- **Business and competitor mapping**: Visualize the distribution and impact of businesses, competitors, and industry clusters for identifying opportunities or risks, measuring cannibalization, and optimizing service delivery.
- **Custom reporting**: Generate informative and detailed reports and infographics summarizing spatial, demographic, and market data. Reports can be shared with stakeholders or integrated into online presentations, hard-copy proposals, or public-facing websites.
- **Territory design**: Create, balance, and optimize market areas based on population dynamics, business constraints, or existing locations. Built-in parametric models support barriers, such as highways or rivers, controlling compactness and allowing a combination of market factors to determine how and where each territory grows. Business rule–based validation and user assignment logic allow territories to be built for an individual town or up to the national level.
- **Spatial modeling and predictive analysis**: Employ advanced spatial analytics to forecast trends, model scenarios, and evaluate future opportunities using a combination of Esri-curated and your organization's own data.

As you will discover in this book, ArcGIS Business Analyst transforms raw data into meaningful, actionable intelligence, supporting strategic planning, operational efficiency, and risk mitigation. It provides capabilities that can be applied immediately, as well as leading to long-term improvements. Many of the analysis workflows or tools build on each other and are companion capabilities, used together as well as individually, to help shape your decision-making. As we designed and wrote this book, we wanted to ensure that each chapter built on and expanded previous knowledge, but that it could also stand alone as a lesson in how to apply ArcGIS Business Analyst in different software environments and industry sectors at different user experience levels and organizational structures.

## Industry applications: ArcGIS Business Analyst in action

ArcGIS Business Analyst's versatility makes it invaluable and widely used across a broad spectrum of industries. Here's how it is used in key sectors:

### Retail

One of the biggest audiences for ArcGIS Business Analyst, retailers apply spatial analytic tools and Esri's rich data to optimize store locations, analyze customer demographics, and understand spending patterns. By mapping competitors' territory and identifying underserved markets, retailers can make informed decisions about expansion, marketing campaigns, and inventory planning. For example, a national chain might use drive-time analysis to determine if, when, and where to build its next set of stores, whereas a local boutique could profile customers to tailor offerings to its immediate market and succeed against bigger, national brands.

### Health care

Health-care organizations rely on ArcGIS Business Analyst to assess community needs, plan facility locations, and target patient outreach programs. Access to care and improved community outcomes rely heavily on geographic patterns and interactions. GIS allows health-care providers to map and model health indicators, population density, demographics, and provider locations to ensure they meet their network adequacy mission to deliver the right services and care. Hospitals and clinics can analyze patient catchment areas and identify gaps in service or discover unidentified needs, supporting operational efficiency and improved patient outcomes.

### Economic development

Government agencies and economic development organizations, tasked with fulfilling the needs and desires of their citizens, use ArcGIS Business Analyst to attract internal investment, identify growth opportunities, and support strategic planning. By visualizing workforce characteristics, business clusters, and infrastructure, governments of all types can communicate the strengths of a region to potential investors. ArcGIS Business Analyst also helps identify workforce development plans; improve transportation networks, housing developments, and community amenities; foster sustainable growth; and improve citizen engagement.

## Architecture, engineering, and construction (AEC)

AEC firms use ArcGIS Business Analyst for site analysis, modeling, and stakeholder engagement. By overlaying demographic and environmental data on potential project sites, AEC firms can assess feasibility, identify risks, and optimize design. ArcGIS Business Analyst supports collaboration by providing clear, data-driven visuals, dashboards, reports, and impact statements for clients and regulators, streamlining approvals and reducing project uncertainty.

## Banking and financial services

Banks and financial institutions use ArcGIS Business Analyst to drive customer services and branch network strategies. ArcGIS Business Analyst enables banks, credit unions, and other financial organizations to evaluate branch locations, understand customer profiles, and assess market potential. By analyzing demographic and economic data, banks can ensure that they comply with financial regulations, target products and services appropriately, mitigate risk, and plan for expansion. The system also supports compliance with regulatory requirements by providing abundant data for community reinvestment and fair lending assessments.

**Table 1. How ArcGIS Business Analyst–derived location intelligence drives business value**

| Business area | Primary location-driven benefits |
|---|---|
| Market planning and real estate | Smarter site selection, optimal store coverage, reduced expansion risk |
| Customer analytics | Better customer understanding, targeted campaigns, higher ROI |
| Merchandising | Localized assortments, improved category management, fewer stock issues |
| Marketing | Targeting the right customers, improved campaign ROI, reduced customer churn |
| Supply chain and distribution | Network organization, optimized warehouse and distribution center placement, balanced production by market |
| Performance and analytics | Key performance indicators (KPIs) and sales performance visualized by location and product |

## Expanding possibilities: Additional use cases

Beyond these core industries, ArcGIS Business Analyst has a long history of applications in education, real estate, insurance, hospitality, and more. Universities use it to plan campus expansions and analyze student populations. Real estate professionals evaluate property values and neighborhood characteristics, supporting site selection for other user types and industries. Insurers assess risk based on geographic and demographic factors. Hotels and travel companies analyze tourist trends and optimize marketing strategies. Public safety organizations plan fire, police, and ambulance services to meet mandatory response times and use ArcGIS Business Analyst extensively in real-time disaster response and emergency situations.

The common thread across these diverse applications is the use of location-based analysis. We have used many cross-industry examples in this book to demonstrate the power of combining spatial data with location intelligence workflows, but we could not include all industries. Each chapter provides insights into how organizations can uncover patterns, anticipate changes, and make strategic decisions using ArcGIS Business Analyst in key markets. As you use this book, we hope you can draw parallels with, and inspiration for, your own organizational needs and personal goals.

## Getting started: Implementation, expansion, and integration

Adopting ArcGIS Business Analyst is straightforward, thanks to its flexible deployment options, license types, and integration capabilities. Individuals and organizations can start with Business Analyst Web App because it offers quick wins and minimal organizational overhead. Business Analyst Web App introduces you to rich content and many of the powerful analytic features, reports, and sharing capabilities of ArcGIS Business Analyst. Many organizations have adopted Business Analyst Web App as their primary corporate GIS tool, provisioning thousands of users, while it's also the go-to platform for small organizations and businesses that need spatial analysis but cannot invest in a dedicated GIS professional.

Many organizations expand to Business Analyst Pro for more advanced analysis, alongside Business Analyst Web App. They often store and share data between both software environments, taking advantage of the most appropriate and easy-to-use tools when and where they need them.

Many GIS professionals who are already familiar with ArcGIS Pro start immediately with Business Analyst Pro because it allows them to expand their capabilities while remaining within a familiar user experience with known workflows.

Larger organizations scale their security and business service integration using ArcGIS Business Analyst™ Enterprise. Business Analyst Pro and Business Analyst Web App users gain enterprise-wide access to other ArcGIS products and applications, business intelligence tools, and corporate data sources while maintaining their own personal data and processing capabilities.

The lessons you learn within this book can be applied to all types of software architectures and implementation approaches. Tutorials written for ArcGIS Business Analyst Web App can be used in both online and enterprise environments.

Esri provides extensive resources—including training, support, and community forums—to help users make the most of ArcGIS Business Analyst. Whether you're new to GIS or an experienced analyst, you'll find the tools and guidance needed to unlock the full potential of location-based business intelligence.

## Conclusion

ArcGIS Business Analyst is a transformative solution for organizations seeking to harness the advantages of location intelligence. We hope that by applying the lessons contained within this book, you can apply the rich data, strong analytics, powerful reporting, and detailed insights offered by ArcGIS Business Analyst in your daily work. By bridging the gap between spatial analysis and business strategy, ArcGIS Business Analyst empowers users to make smarter decisions, drive growth, and create lasting value in a rapidly changing world. This book hopes to further that along and get you started by briefing you on the top 20 essential skills to help you achieve those goals.

# HOW TO USE THIS BOOK

## About this book

*Top 20 Essential Skills for ArcGIS Business Analyst* has been tested for compatibility with ArcGIS Pro 3.7 and the ArcGIS Business Analyst Web App February 2026 update. This book is designed for beginners. No prior experience with ArcGIS Business Analyst is required. The tutorials in each chapter demonstrate a workflow in a hands-on environment, and most should take about 45 minutes to complete.

## Software requirements and licensing

To do the tutorials in this book, you'll need the following: ArcGIS Pro 3.7 and a web browser to access ArcGIS Online. Earlier software versions may not be fully compatible with the tutorial data and may not operate as described.

Hardware requirements for ArcGIS Pro are available at link.esri.com/EsriPress/ProSysReqs.

Information on software access can be found at esri.com.

## Downloading the tutorial data

At the beginning of chapters 17, 18, 19, and 20, a link to the chapter's data is given along with setup instructions. Each chapter's dataset is provided as a zipfile, which should be unzipped and stored in an enclosing folder: C:\Top20BusinessAnalyst.

Tutorial data that accompanies this book is covered by a license agreement that stipulates the terms of use, at link.esri.com/EsriPress/LicenseAgreement.

## Resources, feedback, and updates

The ArcGIS Business Analyst documentation provides comprehensive descriptions of software concepts and tools at link.esri.com/Top20BA/Doc. The Frequently Asked Questions page is a resource for answers to common questions you may have at link.esri.com/Top20BA/FAQ.

Feedback, updates, and collaboration are available at Esri Community, the global community of Esri users. Post any questions or comments about this book to the Esri Press community at link.esri.com/EsriPress/EPCommunity or reach out to Esri Press at esripress@esri.com.

Visit the book's web page for additional information at link.esri.com/Top20BA/Book.

# CHAPTER 1

# Getting started with ArcGIS Business Analyst

## Objectives

- Explore the Business Analyst Web App and Business Analyst Pro interfaces and workflows.
- Create your first Business Analyst Web App project.
- Use the guided tour to explore the interface and toolset.
- Download and configure Business Analyst Pro.

## Introduction

This chapter introduces the Business Analyst Web App and Business Analyst Pro user interfaces and capabilities. As discussed in the introduction, ArcGIS Business Analyst combines software and data with workflows, visualization, and analysis tools to assess sites, understand customers, and generate market insights for locations worldwide. Business Analyst Web App has been designed for the non-GIS user, with different workflows and functionality organized under tabs and side panels. Business Analyst Pro is an ArcGIS extension, which can be found on the Analysis tab and uses the same interface design and layout as other ArcGIS Pro tools and extensions.

Data for more than 170 countries is hosted in ArcGIS Online and is available to Business Analyst Web App through a data source selector or through portal connections in Business Analyst Pro. On-premises data packages are also available for selected countries for Business Analyst Pro. Data packages contain the same data as hosted in ArcGIS Online but do not incur credit cost when performing analysis or creating reports and infographics.

Swapping countries is achieved by setting the country data source in both Business Analyst Web App and Business Analyst Pro.

The following tutorials will introduce you to Business Analyst Web App and Business Analyst Pro for the first time, discuss the interface components, and provide some introductory information for workflows that you will use throughout this book.

# Tutorial 1-1: Access Business Analyst Web App for the first time

In this tutorial, you'll access ArcGIS Business Analyst Web App for the first time. The web app is made available by your organizational administrator, who will assign a license. Once assigned, you can sign in to ArcGIS Business Analyst Web App.

## Start Business Analyst Web App and create a new project

1. In your browser, navigate to bao.arcgis.com.

2. Enter your ArcGIS credentials and click **Sign In**.

> **Note:** You will need to set your organizational URL that is associated with your username when you sign in for the first time. You will not need to do this for subsequent sign-ins.

## Explore the Welcome page and Explore an area

Business Analyst Web App provides a helper screen to showcase the types of workflows, maps, and reports that you can create.

3. On the **Welcome** page, in the **Explore an area** section, click **San Diego Convention Center** to narrow your area of interest.

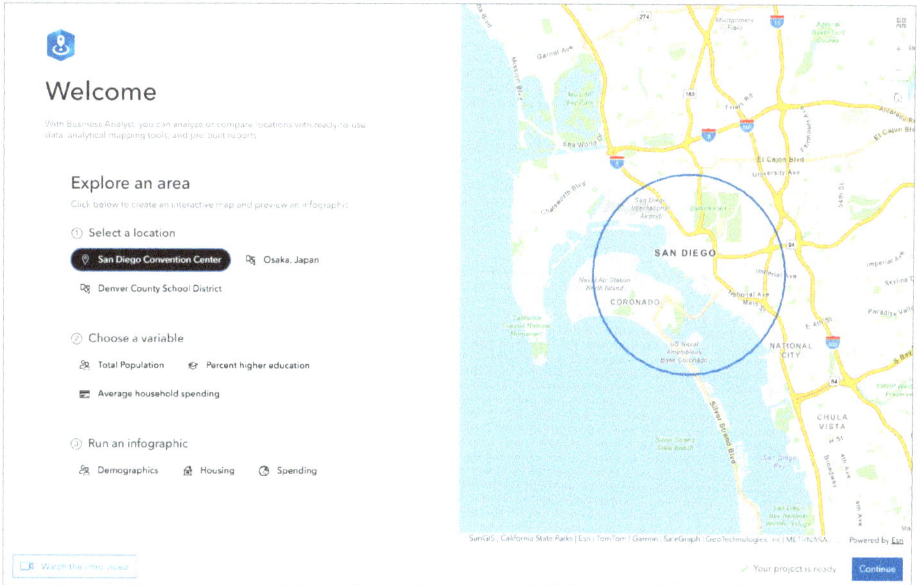


**Note:** You can watch an introductory video at any time when using the **Welcome** screen. The introductory video is also available in the help resources, addressed later in this chapter.

4. In the **Choose a variable** section, click **Total Population**.

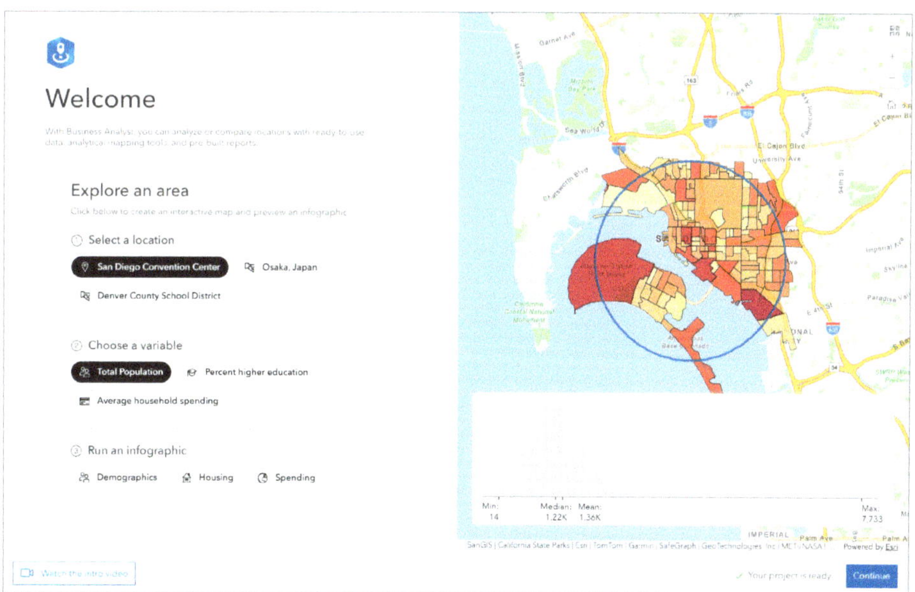

You will see a color-coded map with population, where areas in red have the highest population. You can also click on bars in the histogram to see where census tracts with different population levels are located.

5. In the **Run an infographic** section, click the **Demographics** option.

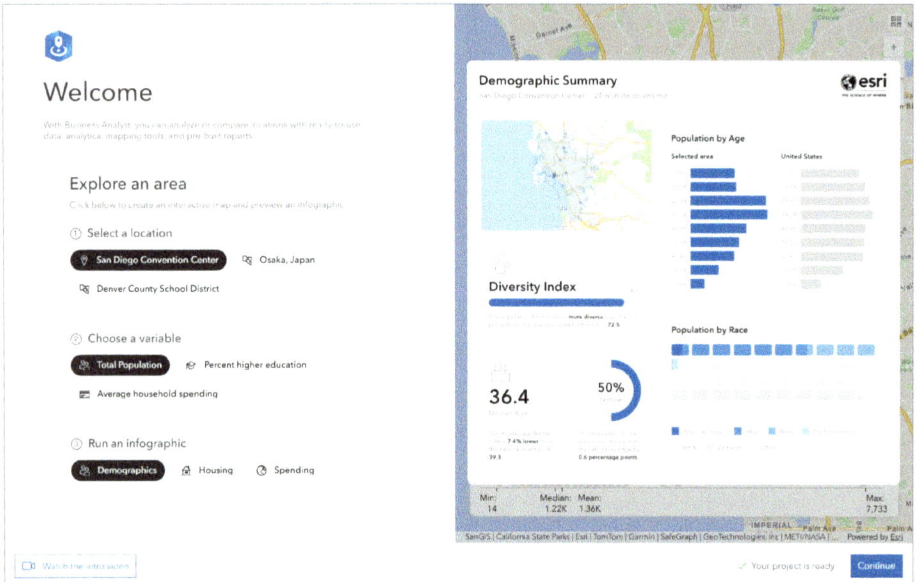


You will see a summary of the demographics in the trade area. This is an example of more than 30 infographic templates available for the United States.

6. When you are finished, click **Continue** to create your project and enter the **Home** page.

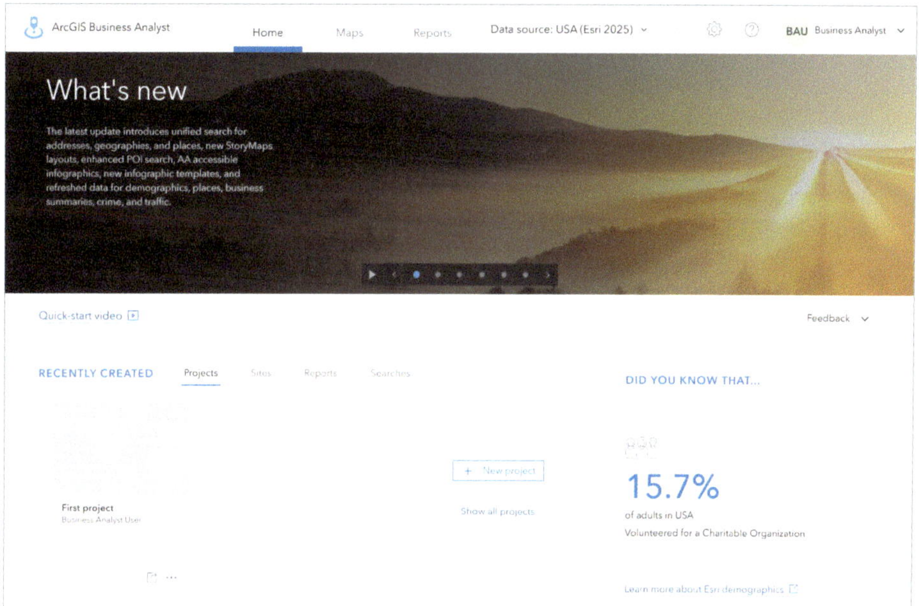


Your first project is now ready. By default, the newly created project is called "First project." You will change the name to better describe the project.

7. On the project card, click the **Options** button (three dots) and click **Edit name**. Change the project name to Ch 1 - Intro to BA Web App.
8. Hover over the map section of the project card and click **Open project** to access the mapping, reporting, and analysis workflows.

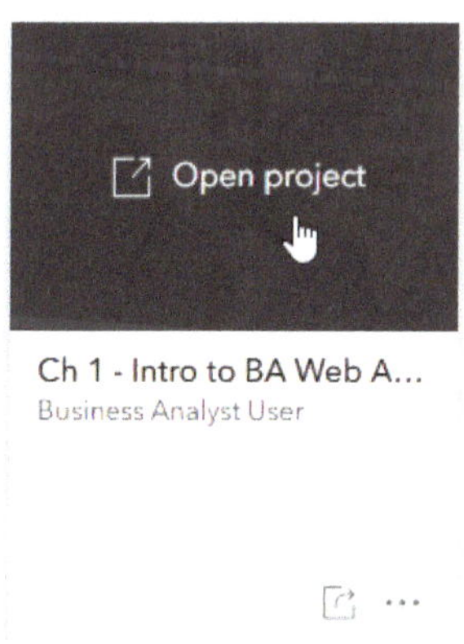

You will now be presented with a map of the United States and the project pane for your first project.

## Explore the Business Analyst Web App user interface

Now that you've created your first project, you can start using the workflows and tools. The following chapters will introduce you to the core capabilities and most important tools. Before you begin working with Business Analyst Web App, it is important to understand the user interface and how the tools are grouped and accessed.

| Label | Name | Description |
|---|---|---|
| 1 | All projects menu | Opens, creates, and manages projects and project items. |
| 2 | Project pane | All content in a project is stored in this pane. The pane expands over the map and any workflows currently open. |
| 3 | Add data | Opens the Add data workflow menu to access web maps and layers, import files, or set up custom data configurations for user-defined datasets. |

| Label | Name | Description |
|---|---|---|
| 4 | Create maps | Opens the Create maps workflow menu to use the color-coded maps, smart map search, and points of interest search workflows. |
| 5 | Define areas | Opens the Define areas workflow menu to find locations, select geographies, and draw polygons. |
| 6 | Run analysis | Opens the Run analysis workflow menu to perform benchmark comparisons, suitability and nearby analysis, examine map gaps (voids), and create threshold areas. |
| 7 | Share results | Opens the Share results workflow menu to create a PDF or image of the current map, as well as share your work in ArcGIS Online, ArcGIS Dashboards, or ArcGIS StoryMaps. |
| 8 | Search bar | Search for addresses, geographies, and places, as well as browse recent searches. |
| 9 | Basemaps button | Provides access to the background maps that provide visual context for the analysis and data you create. |
| 10 | Suggested zoom levels/ Previous zoom | Provides a preconfigured set of geographic zoom levels for managing the scale of the map and the extent of the map window. Previous zoom extent controls allow you to move between prior pans and zooms. |
| 11 | Default extent/Find my location | Resets the map extent to the user-defined default. Find my location zooms to the user's current IP location if location sharing is enabled. |
| 12 | Hide interface | Removes all panes and workflow tabs to allow the user to focus on specific details or site-specific action, such as infographics and comparison analysis. |
| 13 | Map tools | A set of tools to interact with the map, including zoom, add annotation, view historical traffic data, add boundaries, and other tools. |
| 14 | Ribbon | Provides access to the Home page, Maps and Reports tabs, data switcher, preferences, and other tools. |
| 15 | Reports tab | Accesses the Reports tab to run, edit, and manage standard, shared, and user-created infographics. |
| 16 | Data source menu | Swaps between country-specific data sources. |

| Label | Name | Description |
|---|---|---|
| 17 | Business Analyst assistant | Launches the Business Analyst assistant, a natural language interface to assist with specific tasks in Business Analyst Web App. |
| 18 | My preferences | Configures and sets software preferences, including the default map extent, preferred basemap, project colors, units, map and report references, and workflow options. |
| 19 | Help | Accesses the integrated help resources, including "What's new" in each release, guided tours, Support contact, and documentation. |
| 20 | Account details | Accesses the user profile and sign-out controls. |

The image and table describe each of the tabs, panes, and toolbars. Most interactions will use the ribbon, the white menu bar at the top of the interface, to access the functional tabs (Home, Maps, Reports) and user configuration options.

The Project pane is used to manage, organize, and open project content. Map controls, which are in the map frame, allow you to change the underlying basemap and configure the zoom scale and what level of detail the map shows.

> **Note:** You can set the default map extent in user preferences to change it to a specific area of interest, such as the city of San Diego or Dallas County. This allows you to use the **Default extent** button to return to the same area each time.

The map tools contain a range of tools to add annotation to the map, generate a quick output of the map window, share data with ArcGIS Online and other functions, including adding historical traffic patterns, geographic boundaries, and labels to the map.

## Explore the project pane

The project pane is used to access sites, imported maps, previously run reports, and analysis layers. It operates as a table of contents for all the maps you save and reports that you run and saves data that you import.

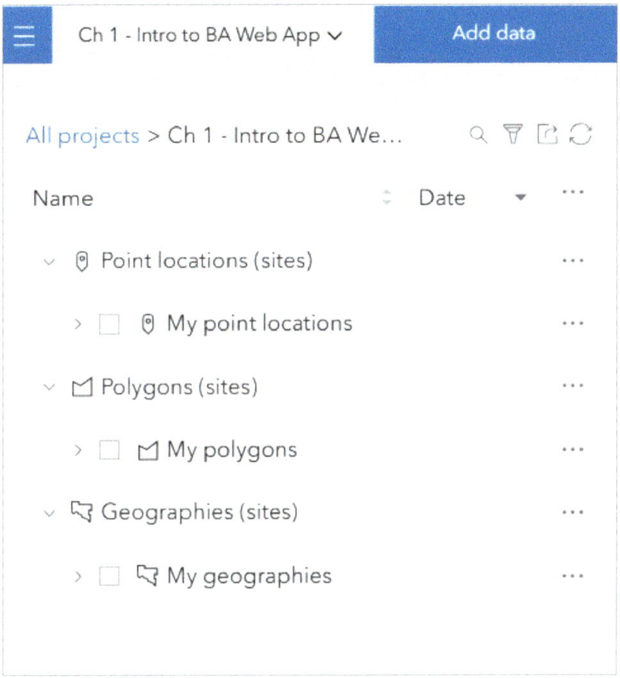


Site layers are organized into three distinct types:

- Point layers, such as an address or place
- Polygons that you draw
- Geographies, such as San Diego County or Colorado

Each time you create a new site, it is stored in the appropriate site type. As you will learn in the next chapter, "Generating Trade Areas," a point site is based on some type of distance metric, such as ring distance, drive time or distance, or a walk time.

Geographies are created using the Define areas workflow tab using either the Select Geography or Draw Polygon tools. When selecting a geography, you are presented with the different geography types, such as counties, zip codes, census tracts, and designated market areas (DMAs) in the United States. Geographies can be built by combining multiple areas into a single geographic area.

As you use Business Analyst Web App more, you will create additional types of data as shown in the accompanying screenshot. Some countries support hexagon geographies, which have a common size at each scale. Hexagons, also known as hexbins, are a popular geographic unit for spatial and data scientists because they are the same size at any scale.

Analysis tools, such as threshold areas, suitability analysis, and benchmark comparisons, allow you to save maps to a project so that you can reuse them in other analysis workflows. You can also save your color-coded maps and smart map searches, so that you can return later to reexamine your maps and provide additional context for analysis.

Points of interest searches allow you to save a collection of points, such as all the coffee shops in the area, for later use in analysis workflows, such as threshold areas, suitability analysis, and benchmark comparisons, or they can be used as individual sites.

Web maps and layers from ArcGIS Online or shared within your organization are also added to the project contents. This allows you to take advantage of the integration with ArcGIS Living Atlas of the World, as well as other data that might be shared with you.

Finally, when you run and save reports, a copy is stored locally in the project and linked to the site on which it was run, so that you can minimize the number of times you create duplicate reports, saving time and credits.

As you become more familiar with the project content and the project concept, you might want to create different projects for different areas that you analyze or different project themes, such as risk analysis, site selection, population modeling, or other topics. We encourage you to create a new project for each chapter so that you can see how data is created and managed in Business Analyst Web App.

Now that you understand more about Business Analyst Web App, we encourage you to explore the functionality and capabilities in the next chapters.

# Tutorial 1-2: Configure Business Analyst Pro for the first time

In this tutorial, you'll learn how to access and start Business Analyst Pro for the first time. To access Business Analyst Pro, you must download ArcGIS Pro through your organization, install the program, and ensure that your administrator has assigned you a license for the ArcGIS Business Analyst extension.

If you cannot start ArcGIS Business Analyst or do not have the right privileges, use the links to the FAQs and help documentation pages listed in the "How to Use This Book" section of the book.

## Create a project

You will start by creating a new project in ArcGIS Pro.

1. Start ArcGIS Pro. If prompted, sign in using your licensed ArcGIS organizational account.
2. Under **New Project**, click **Map**.
3. In the **New Project** dialog box, apply the following settings:
   - **Name**: Ch 1 - First project
   - **Location**: C:\Top20BusinessAnalyst

   **Note:** This book uses **C:\Top20BusinessAnalyst** for the folder location, but you can use any folder.

4. Click **OK** to create your project.

   The ArcGIS Pro interface appears.

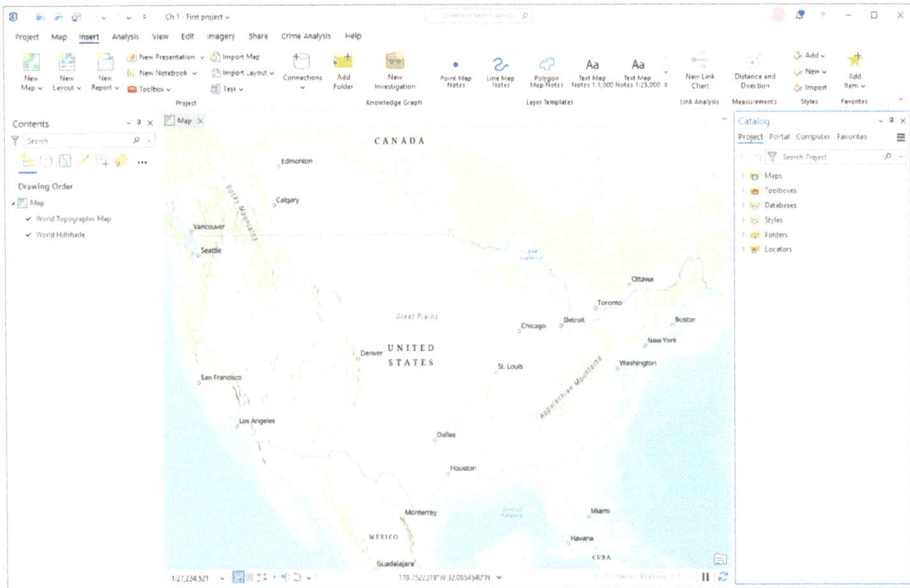


## Explore the Business Analysis gallery

The Business Analyst Pro extension tools are available on the Analysis tab.

5. At the top of the interface, on the ribbon, click the **Analysis** tab.

6. On the **Analysis** tab, in the **Workflows** group, locate the **Business Analysis** button.

If you are fully licensed, the icon will be clickable.

7. Click **Business Analysis** to open the gallery of Business Analyst Pro workflows.

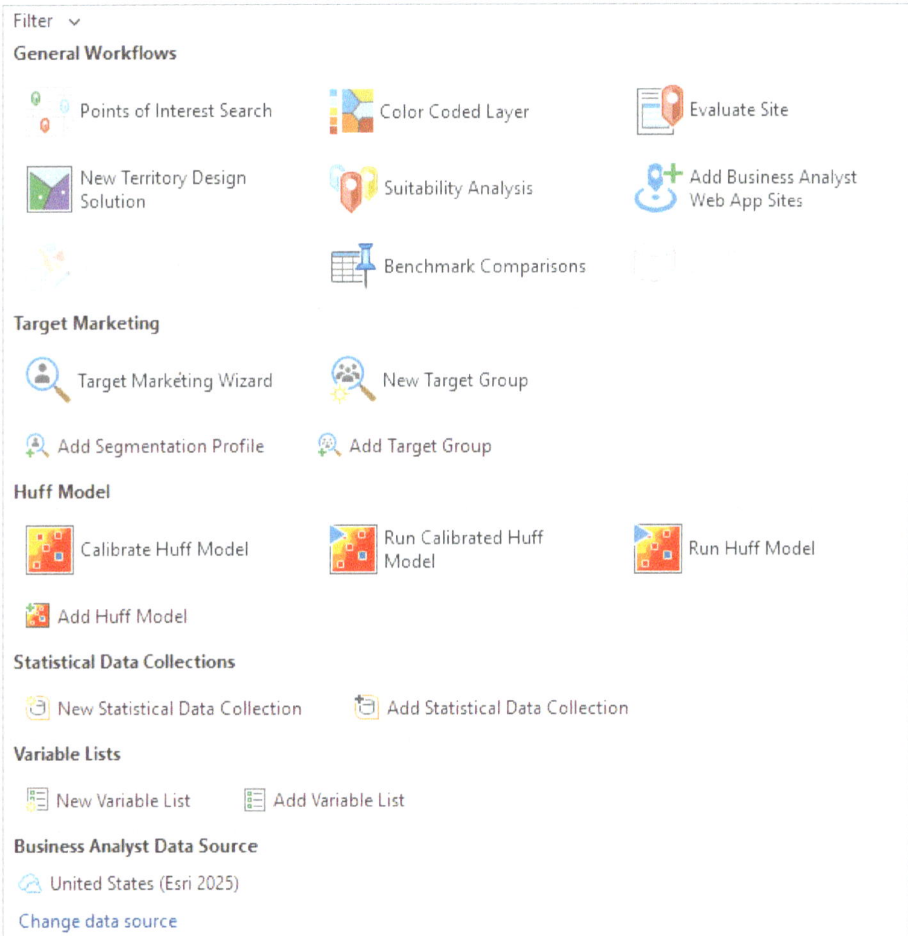


The Business Analysis gallery includes a curated set of workflows, including common analyses, setting the data source, and creating custom data. The available tools are based on the type of Business Analyst data source you are using. Many of the workflows in the gallery are entry points to common geoprocessing tools.

**Note:** You can click the **Filter** option in the gallery to hide or show individual categories.

## Explore Business Analyst Pro through the Geoprocessing pane

You can also access Business Analyst Pro tools and workflows through the Geoprocessing pane.

8. On the ribbon, make sure you are on the **Analysis** tab. In the **Geoprocessing** group, click **Tools** to open the **Geoprocessing** pane.

9. In the **Geoprocessing** pane, click the **Toolboxes** tab. Locate and expand the **Business Analyst Tools** toolbox.

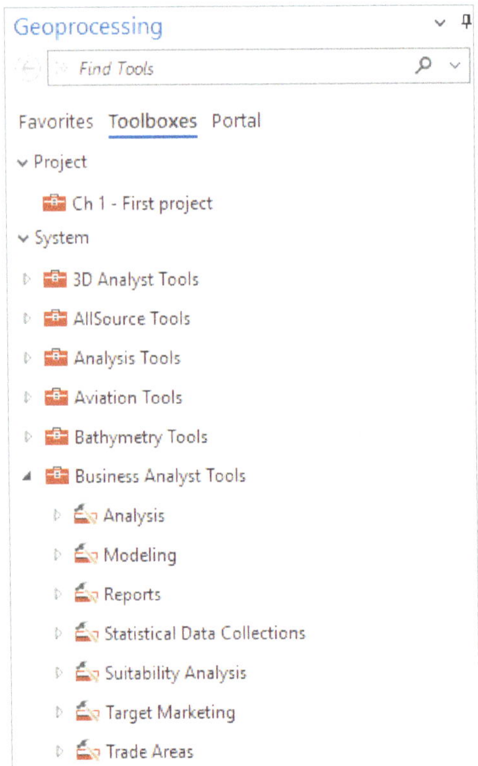


10. Expand any toolset, such as the Suitability Analysis toolset, to explore its contents.

11. On the **Toolboxes** tab, scroll down the list of toolboxes and expand the **Territory Design Tools** toolbox to explore the following toolsets:

- **Analysis**
- **Territory Solution**

**Note:** You can find the specific tool you want to use by using the **Find Tools** search field in the **Geoprocessing** pane.

Chapters that feature Business Analyst Pro access tools and workflows through a combination of the Business Analysis gallery and the Geoprocessing pane.

## Explore Business Analyst Pro data sources

Depending on your organizational setup, location, and data access rights, you should automatically see a geographic area, known as the Business Analyst data source, when accessing tools from the Business Analysis gallery.

In the example shown, the data source is set to an ArcGIS Online hosted data source, United States (Esri 2025).

In the next example below, the data source is set to a local dataset (which requires an additional purchase from your organization), USA_ESRI_2025.

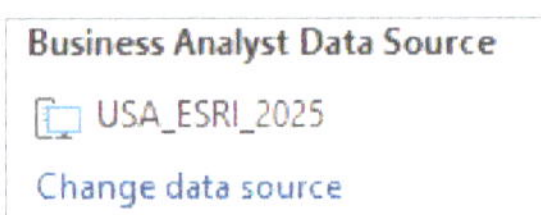


*Important: If the local data is not downloaded and connected, you will not be able to complete the tutorials in the target marketing or statistical data collection chapters.*

You can change the data source to use the same data that is available in Business Analyst Web App.

12. Click **Change data source** to bring up the **Business Analyst Data Source** window.

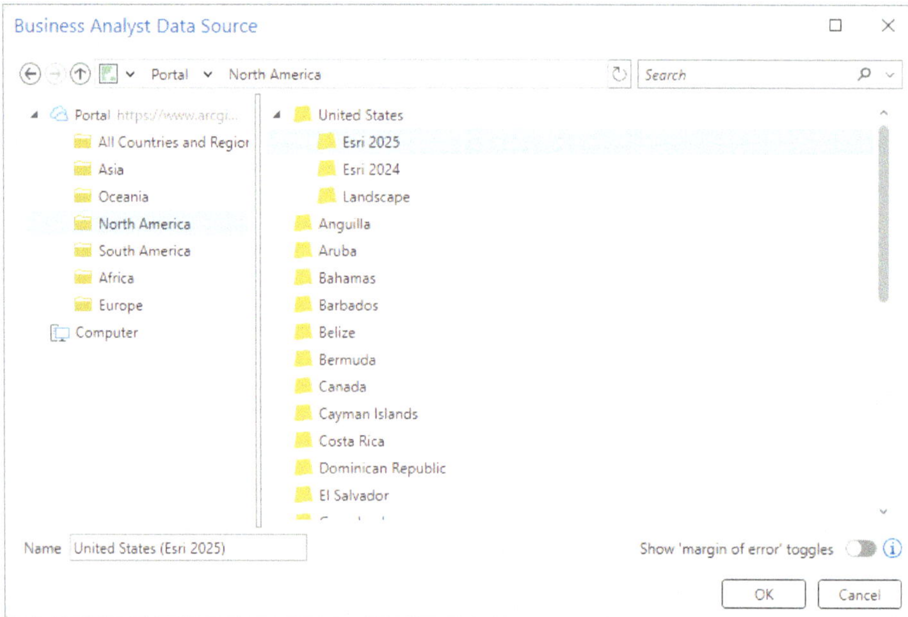


The default country dataset is determined by your computer's regional settings. You will change the data source to Italy.

13. Under **Portal**, click **Europe** > **Italy** and then click **OK.**

You will now see Italy as the Business Analyst data source. The cloud icon signifies that this is an online data source. To change the data source back to your default country, repeat the above actions and select your country.

14. Change the data source to **United States (Esri 2025)**.

# Summary

In this chapter, you were introduced to the Business Analyst Web App and Business Analyst Pro user interfaces. Based on this knowledge, you can proceed to the next chapters prepared to complete the tutorials and understand where to find the tools and functionality described in each tutorial.

## Workflow

1. Follow the ArcGIS Business Analyst Web App welcome tutorial.
2. In ArcGIS Business Analyst Web App, click the buttons and tabs to familiarize yourself with the interface.
3. In ArcGIS Pro, create a new project.
4. Explore the Business Analyst Pro extension toolbox.
5. Change the Business Analyst data source to match the workflow you will be doing.

# CHAPTER 2

# Generating trade areas

## Objectives

- Create a site in ArcGIS Business Analyst Web App.
- Generate and analyze trade areas in ArcGIS Business Analyst Pro.

## Introduction

In ArcGIS Business Analyst, a site is an area where analysis is performed. You can create a site in Business Analyst Web App in several ways, including placing a pin on the map, choosing from standard geography levels, or drawing a polygon. With Business Analyst Pro, you can create sites using geoprocessing tools in the Trade Areas toolset.

You can choose how the area around a site is measured using rings, drive times, or walk times. In the ArcGIS Pro extension, these areas are called trade areas. Typically, geography, distance, and time-based trade areas are the most used methods for generating trade areas. They can be defined by straight-line distance or time, network distance or time, or administrative boundaries.

# Tutorial 2-1: Create a site in Business Analyst Web App

With Business Analyst Web App, you can create a site in several ways. For this tutorial, you will place a point on the map using the Pin map tool.

## Create a project

You'll first log in to Business Analyst Web App and create a new project.

1. In your browser, navigate to bao.arcgis.com.
2. Enter your ArcGIS credentials and click **Sign In**.
3. On the **Home** tab, under **Projects**, click **New project**.
4. In the **Create project** window, for **Project name**, type Ch 2 - Trade areas. Click **Create**.
5. After the project is created, hover over the project card and click **Open project**.

## Place a point on the map

One of the ways that you can create a site is by placing a point on the map. Use the Pin map tool to get started.

6. On the right of the map, in the map tools, click the **Pin** tool.
7. Click a location on the map, such as Denver, Colorado, United States.

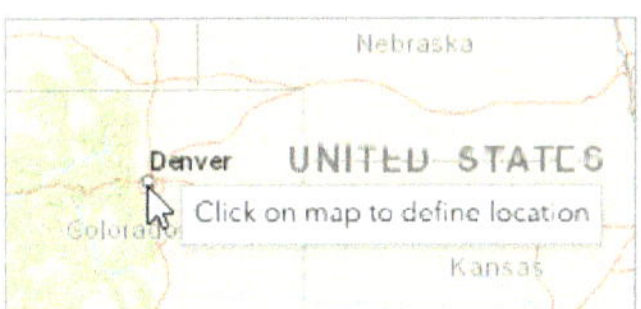


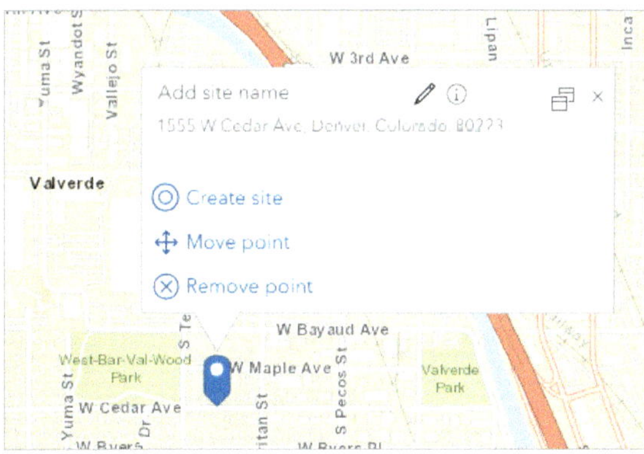


The pin is placed at the point you selected.

> **Tip:** There are two other ways that you can place a point on the map. You can type an address or latitude coordinates in the application search bar. You can also click the **Define areas** tab, click **Find location**, and then enter an address or coordinates.

8. In the pop-up above the pin, click **Create site**.

   Now you can choose how the area around the pin is measured. There are three options. Rings are drawn around the point using specified distances. The drive time represents the distance traveled to or from the site in a specified time frame. The walk time represents the specified times or distances based on paths and roads that allow pedestrian traffic.

   By default, the site uses rings with a radius of 1, 3, and 5 miles.

> **Tip:** The unit measure depends on your data source. You can use nonwhole numbers, such as .1, .3, and .5, for ring sizes. The limit for ring size is 1,000 miles or kilometers.

9. With the default values, click **Apply**.

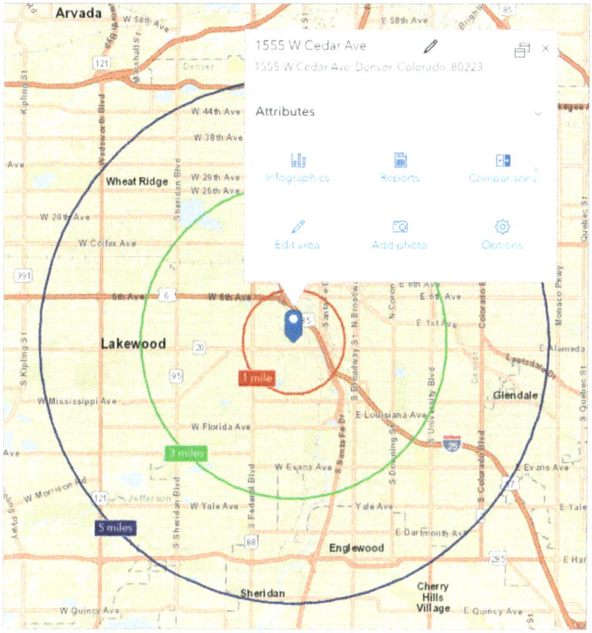


Your site is saved to your project.

## Take it to the next level

Once you've created your site, you can use its pop-up menu to work with it. For example, you can edit the name of the site, provide site attributes, run infographics and reports, and run geography comparisons.

Geography comparisons instantly contextualize your site, comparing key demographic variables for your site and its nested geographies, such as the local block group or census tract, county, state, and country. The geography levels depend on where your site is located, such as using provinces in Canada or districts in India.

# Tutorial 2-2: Generate trade areas in Business Analyst Pro

In Business Analyst Pro, trade areas represent locations where you want to perform analysis, such as a ring or travel time around a point location, a geographic boundary, such as a county or block group, or a polygon shaped by the locations of your customers or constituents. In this tutorial, you will map trade areas around grocery stores and then measure the amount they overlap.

## Create a project

1. Start ArcGIS Pro and sign in using your licensed ArcGIS organizational account.
2. Under **New Project**, click **Map**.
3. In the **New Project** dialog box, for **Name**, type Ch 2 - Grocery store trade areas. Save the project to **C:\Top20BusinessAnalyst**.

> **Tip:** It is helpful to create a folder on your C: drive titled **Top20BusinessAnalyst** to organize your projects for this book.

## Perform a points of interest search

4. Zoom the map to the general area of Chicago, Illinois.

5. On the ribbon, click the **Analysis** tab.

6. In the **Workflows** group, click **Business Analysis** to open the gallery of tools. Under **General Workflows**, click **Points of Interest Search**.

   The Points of Interest Search pane appears.

7. For **Step One: Initial Search**, under **Name or keyword**, type Trader Joe's and click **Next**.

8. For **Step Two: Filter**, beneath the results list, click **Preview Map** to show all the Trader Joe's stores the search returned.

9. In the results list, expand **City**. Hover next to **Chicago** and click **Only**.

10. For **Output**, change the name of the layer to Trader_Joes_Chicago.

11. Click **Finish**.

    The selected locations are saved to a new layer, visible on the map and in the Contents pane.

## Generate trade areas for stores

12. On the ribbon, click the **Analysis** tab. In the **Geoprocessing** group, click **Tools**.
13. In the **Geoprocessing** pane, search for Generate Drive Time Trade Areas and then click the tool to open it.
14. In the **Generate Drive Time Trade Areas** tool, apply the following settings:
    - For **Input Features**, select **Trader_Joes_Chicago**.
    - For **Output Feature Class**, change the name to Trader_Joes_Chicago_drivetimes.
    - For **Distance Type**, leave the default, **Driving Time**.
    - For **Input Method**, leave the default, **Values**.
    - For **Distances**, type 5.
    - For **Distance Units**, leave the default, **Minutes**.
    - For **ID Field**, select **Esri PID**.
    - For **Dissolve Option**, leave the default, **Overlap**.
15. Click **Run**.

**Tip:** If prompted to estimate credits, at the top of the tool, click **estimate credits**. The tool is estimated to consume about 3 credits.

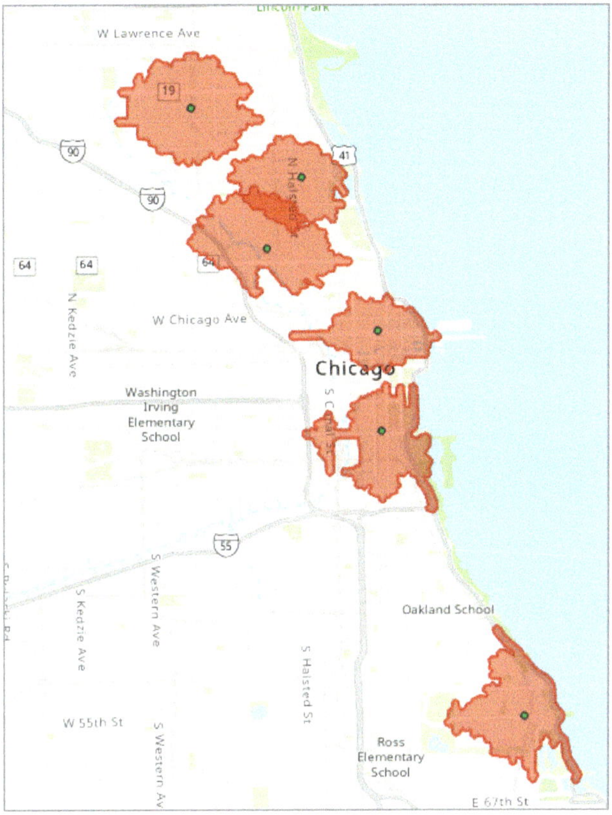


The tool runs, creating polygons that show 5-minute drive times around each Trader Joe's store. There are two stores in the north of the city where the trade areas overlap.

## Measure trade area overlap

Now you'll learn how much they overlap.

16. On the map, click the polygons of the overlapping points to view their pop-ups. Note the value in the **Object ID** field, such as **4** or **6**. Close the pop-ups.

17. At the top of the **Generate Drive Time Trade Areas** tool, click the back button to return to the **Geoprocessing** pane.

18. In the search field of the **Geoprocessing** pane, search for Measure Cannibalization and then click the tool to open it.

19. In the **Measure Cannibalization** pane, apply the following settings:

    - For **Input Features**, click the **Trader_Joes_Chicago_drivetimes** layer.
    - For **Trade Area ID**, click **Street**.
    - For **Trade Area Description**, click **Street**.
    - For **Output Feature Class**, change the name to Trader_Joes_Chicago_drivetimes_overlap.
    - For **Store ID Field**, click **Street**.
    - Check the box for **Create Report**.

20. Click **Run**.

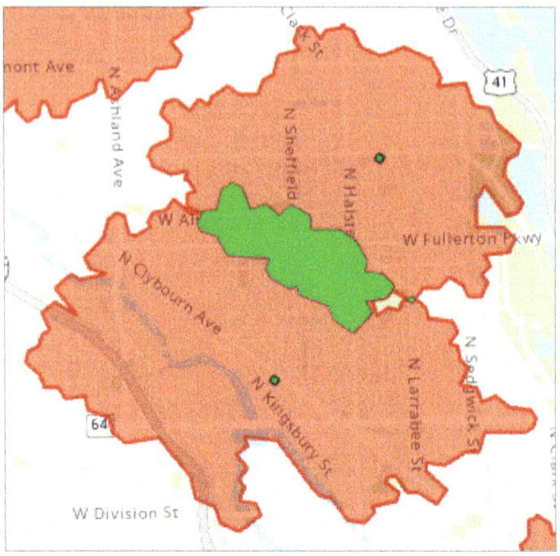


The tool runs and the overlapping area is mapped as a new layer, visible in the Contents pane.

21. Click the overlap polygon to view its pop-up, which shows that about 7% of the trade areas are overlapping. Close the pop-up.

### Take it to the next level

You can use the **Measure Cannibalization** report to learn more about each trade area and the potential effect of the overlap.

1. At the bottom of the **Measure Cannibalization** pane, hover over the **Measure Cannibalization completed** message.
2. Next to **Output Report**, click the link to open the report.

The trade area around the store on N. Clybourn Ave. is about 40% larger than the trade area around the store on W. Diversey. Although the W. Diversey store could potentially cannibalize about 12% of the market for N. Clybourn, the impact on N. Clybourn could be larger, at about 16.5%. Understanding the relationship between these closely sited stores can help decision-makers allocate resources and market to each store's customers while minimizing the effects on each store's sales.

## Summary

In this chapter, you used Business Analyst Web App to create a site around a point on the map. You also used trade area tools in Business Analyst Pro to map the market for a grocery store chain and visualize the overlap of the two stores' trade areas.

### Workflow

1. Determine the locations you want to analyze.
2. Create trade areas around the points.
3. Analyze the relationship between the trade areas.

# CHAPTER 3

# Creating color-coded maps

## Objectives

- Use areas of interest and level of detail settings to set up a map.
- Experiment with color schemes.
- Explore the results of a color-coded map.

## Introduction

Color-coded maps, also known as choropleth maps, use colors to represent the distribution of a variable or variables in communities. Using color-coding on a map presents patterns in data.

In ArcGIS Business Analyst, you can map demographic information, consumer spending patterns, information about a business, and other variables. You can use one or two variables in the map; using two variables creates a bivariate map that you can use to compare data patterns.

You can create color-coded maps in Business Analyst Web App and color-coded layers in Business Analyst Pro.

# Tutorial 3-1: Create color-coded maps in Business Analyst Web App

In Business Analyst Web App, your color-coded map is presented in the following ways:

- The variable is added to the map and displayed at the selected geography level.
- The Results pane displays a summary, histogram, and table.

## Create a project

1. In your browser, navigate to bao.arcgis.com.
2. Enter your ArcGIS credentials and click **Sign In**.
3. On the **Home** tab, click **New project**. Name your project Ch 3 - Population in Austin and click **Create**. Open your project.

   **Note:** If you are prompted to clear the map of previous data, click **Yes**.

## Set up your map

The first part of making a color-coded map is setting your geography, specifically the area of interest and level of detail. You can use either standard geographies, such as zip codes, or hexagons for the map.

For this tutorial, you'll be studying Austin, Texas, in a map using hexagons.

4. To start making your color-coded map, on the **Maps** tab, click **Create maps** and click **Color-coded maps**.
5. In the **Color-coded maps** pane, expand the **Geography** section. Click the **Area of interest** search field and type Austin, TX.
6. In the results, select **Austin City, TX (Place)**.

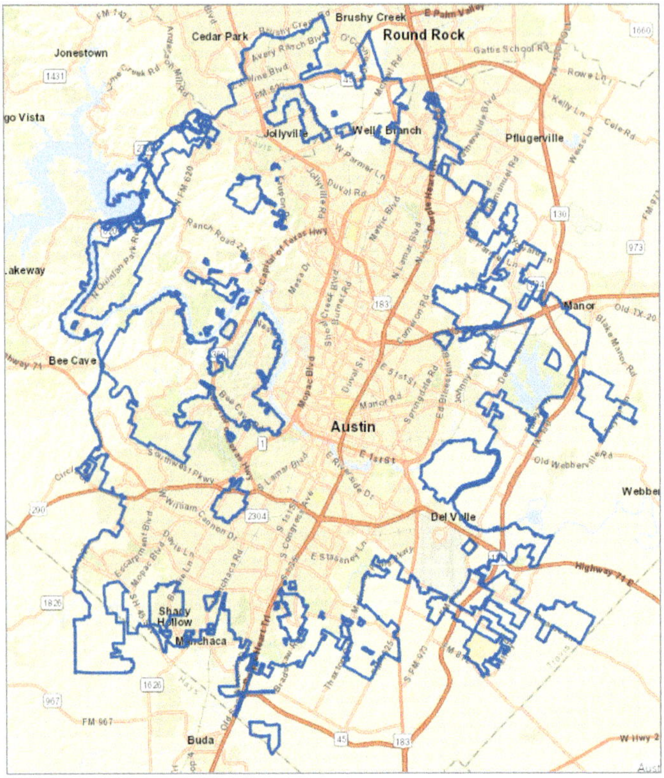


7. Under **Level of detail**, click **Hexagons**.

   By default, the level of detail is set to Resolution 7 hexagons.

   **Tip:** If you're using standard geographies, click **View glossary** to learn more about the definition of each geographic level. If you're using hexagons, the level of detail refers to the hexagon size, known as the resolution. Hexagons can vary in resolution, from 0 (largest) to 15 (smallest). Business Analyst Web App supports resolutions 2 through 7, which are available by zooming in on the map extent.

After setting up the geography of your color-coded map, you will add a variable.

For this tutorial, you'll add the Daytime Population variable to understand the total population in Austin during the day, which accounts for people who may commute to the city for work.

8. In the **Color-coded maps** pane, in the **Variables** section, view the variables under the **Popular** tab. Select the **Daytime Population** variable.

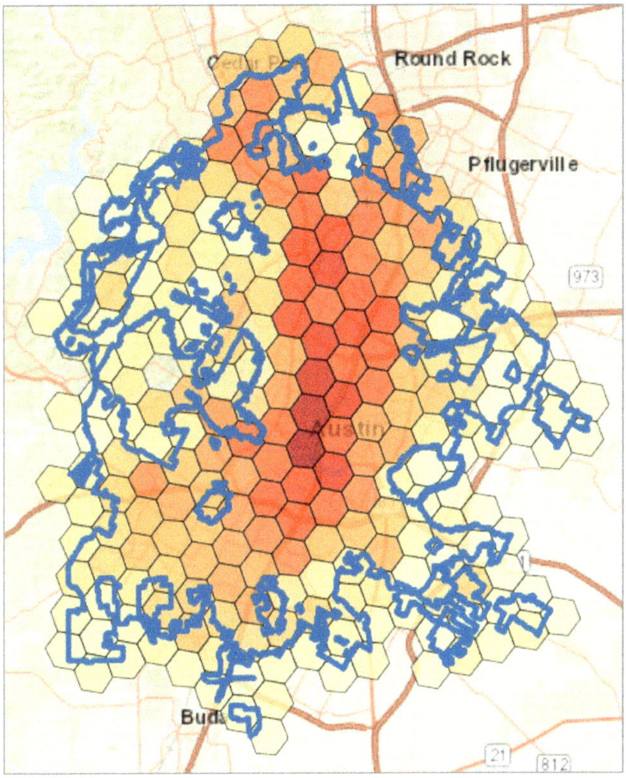


The map is automatically created. The default map style highlights the data variances.

You can visualize your data through a varicty of map styles. When you style a color-coded map, the type of data and number of variables determine the styling options. You can experiment with color ramps, line weights, transparency, symbols, and other graphic elements and see your choices reflected immediately in the map. Not all map styles are available for every color-coded map. Additionally, the map style depends on whether you are using one or two variables.

9. Observe the **Legend** section to see what the color-coding indicates.

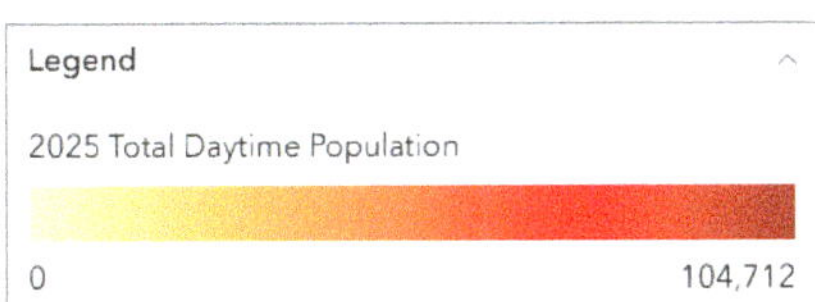

With the default color theme, dark-red hexagons indicate high daytime population, whereas pale-yellow hexagons indicate low daytime population.

**Tip:** It is useful to classify your data to limit the number of colors used on the map and clearly define the values that each class visualizes.

In the **Legend** section, the **Classify data** toggle button is turned on by default. You can modify the number of classes and associated color-coding. By default, five classes are created.

## Explore the results

Below the map is the Results pane. You can explore the color-coded map results as a summary, histogram, and table.

10. In the **Results** pane, on the **Summary** tab, view the summary of the results, including:

    - The number and type of variables you mapped
    - The count of areas, which may be in the summary
    - The average value of the variable calculation
    - The top five and bottom five areas
    - A statistical trends chart
    - A relationship chart

11. Click the **Histogram** tab to view the results for each variable in a histogram.

    **Tip:** Hover over an item in the histogram to highlight the corresponding area on the map.

12. Click the **Table** tab to view the results in a table.

## Take it to the next level

In Business Analyst Web App, use the **Geography** section of the workflow pane to define the analysis extent and level of detail. The two work in parallel, so you can clip the display to focus on a particular area or site.

- The area of interest controls what geographic area is displayed on the map.
- The level of detail controls which geographic hierarchies or areas are used in the map display.

You can use the 10,000-feature limit as a general guideline to understand which combinations of analysis extent and level of detail will cause certain geographies to not be mapped as part of your analysis.

For example, in Brazil, there are just 27 federative units with 315,788 census sectors. On average, each state has 11,696 sectors (315,788 divided by 27), so it is likely that you will run into the area display limit whenever using a state as the analysis extent.

In the United States, there are 239,203 block groups and 50 states, plus the District of Columbia, which equates to 4,690 block groups per state on average. As the size and population of states varies considerably, it is not possible to map block groups for all of California, New York, or many other populous states. In Canada, some of the 13 provinces or territories cannot be mapped at the dissemination area (DA) level because there are 56,590 DAs.

Similarly, in Italy and Portugal, hundreds of thousands of detailed census areas (401,547 in Italy and 210,170 in Portugal) restrict mapping at the district or regional level.

# Tutorial 3-2: Create color-coded layers in Business Analyst Pro

In Business Analyst Pro, your color-coded layer is presented in the following ways:

- The variable is added to the map and displayed at the highest geography level.
- The Contents pane contains a group layer as well as sublayers for each geography level. To view different geography levels on the map, you can zoom in or out or select a sublayer in the Contents pane.
- The selected variable is added to the group layer Symbology pane, under Variable.

## Create a project

1. Start ArcGIS Pro. If prompted, sign in using your licensed ArcGIS organizational account.

2. Under **New Project**, click **Map**.

3. In the **New Project** dialog box, apply the following settings:

   - **Name**: Ch 3 - Color-coded layers
   - **Location**: C:\Top20BusinessAnalyst

   **Tip:** This book uses **C:\Top20BusinessAnalyst** for the folder location, but you can use any folder.

4. Click **OK** to create your project.

## Create a color-coded map

Now, you can start making your color-coded layer for your map.

5. On the ribbon, click the **Analysis** tab. In the **Workflows** group, click **Business Analysis**.

   **Tip:** Make sure that the **Business Analyst Data Source** is set to United States using the most recent release of data, such as Esri 2025.

6. From the **General Workflows** section, select **Color Coded Layer**.

The data browser appears.

7. In the **Data Browser** window, click the search field and type population density.

8. In the results, select **2025 Population Density (Pop per Square Mile) (Esri)**.

9. Click **OK**.

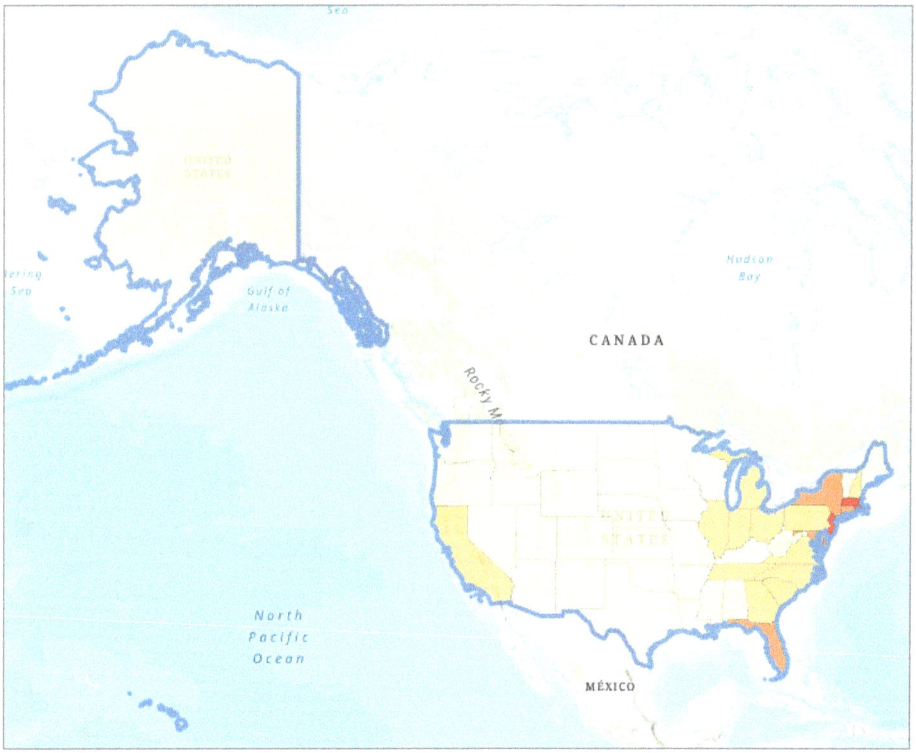


The color-coded map is created. It shows the population density of the United States, with the default color scheme using red to indicate higher population density and yellow to indicate lower population density.

## Modify the map symbology

The Symbology pane appears when you create the color-coded map. You can modify the map symbology so that you have custom map styling. For example, you can change the method used to classify the colors, the number of classes used, and the color scheme.

10. In the **Symbology** pane, under the **Style** section, change the **Method** to **Geometric Interval**.

The geometric interval method creates class breaks based on class intervals that have a geometric series. The geometric coefficient in this classifier can change once (to its inverse) to optimize the class ranges.

11. Click the **Color scheme** drop-down menu and select a different color scheme, such as **Purple-Red Light to Dark**.

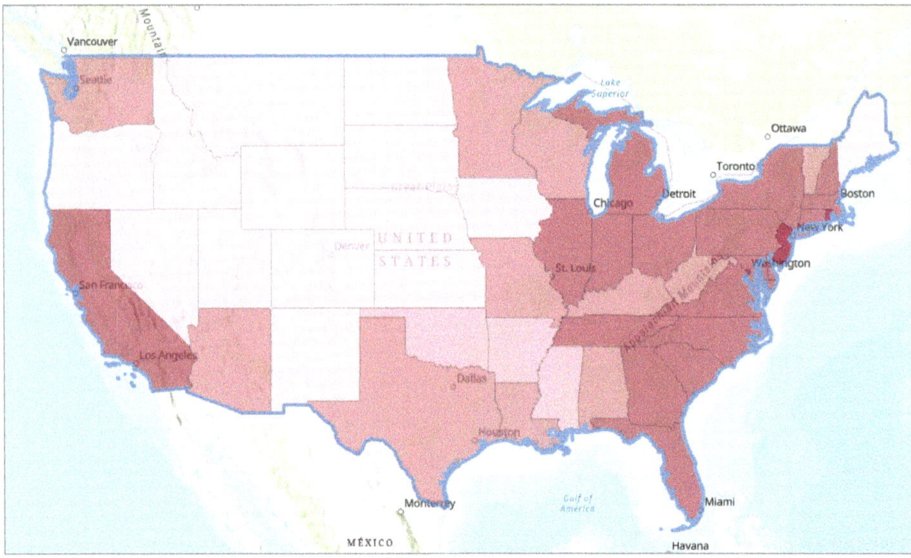


## Set the geography for the analysis

You can also change the area of interest for the color-coded map, the type of boundaries used (standard geographies, such as zip codes or hexagons), and select a level of detail.

12. In the **Symbology** pane, under the **Geography** section, click the **Area of interest** field and type Pittsburgh.

13. In the results, select **Pittsburgh city (4261000)**.

    The map zooms to Pittsburgh, Pennsylvania.

14. For **Boundary mode**, click the **Hexagons** tab.

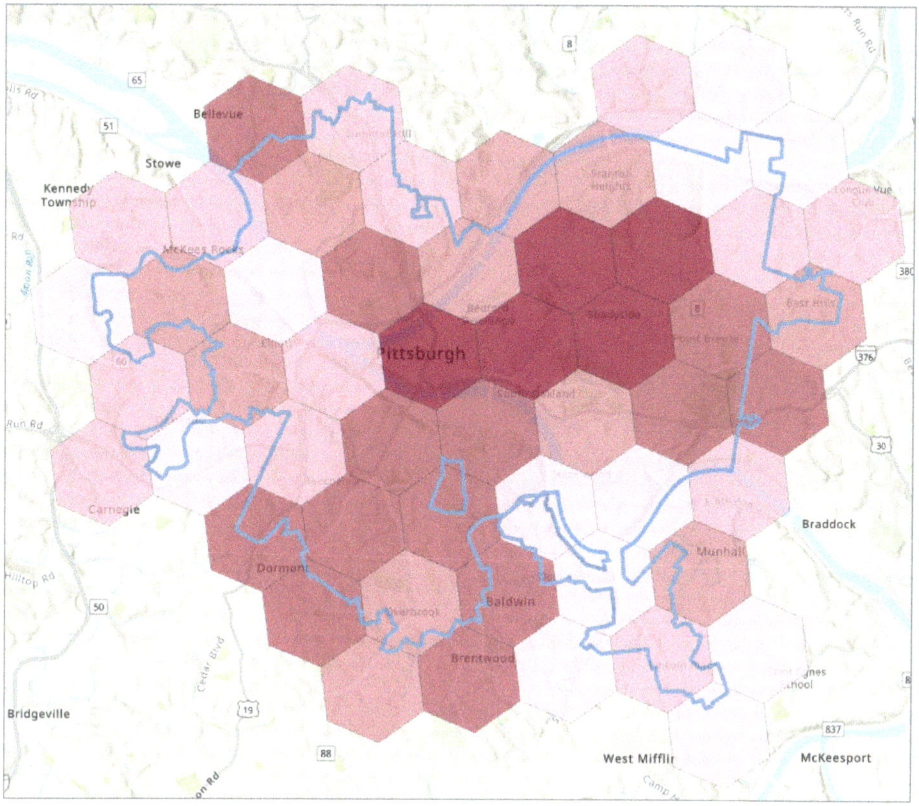


The map adjusts to use hexagons. Switching the boundary mode replaces all existing sublayers in the color-coded group layer.

> **Tip:** Hexagons are available as a geography option for all local datasets, as well as select online datasets. Switching the boundary mode replaces all existing sublayers in the color-coded group layer.

## Explore the results

The results appear on the map and in the Results pane, which features a summary, histogram, and table.

15. At the bottom of the **Symbology** pane, click **View Results**.

    By default, you see the Summary tab first. The summary may include the count of areas, average value of the variable calculation, the top five and bottom five areas, a statistical trends chart, and a relationship chart.

16. View the information on the **Summary** tab.

17. Click the **Histogram** tab to view the results for each variable in a histogram.

18. Click the **Table** tab to view the results in a table.

## Take it to the next level

After you create your color-coded map of population density, you can explore more demographic information about communities that you are interested in. On the ribbon, click **Color Coded Group Layer**. In the **Results** group, click **Infographics**. On the map, click on a hexagon representing a community of your choice. In the **Infographics** window, change the **Template** option to **What's in My Community?** In this infographic, you can learn more about the people living in an area.

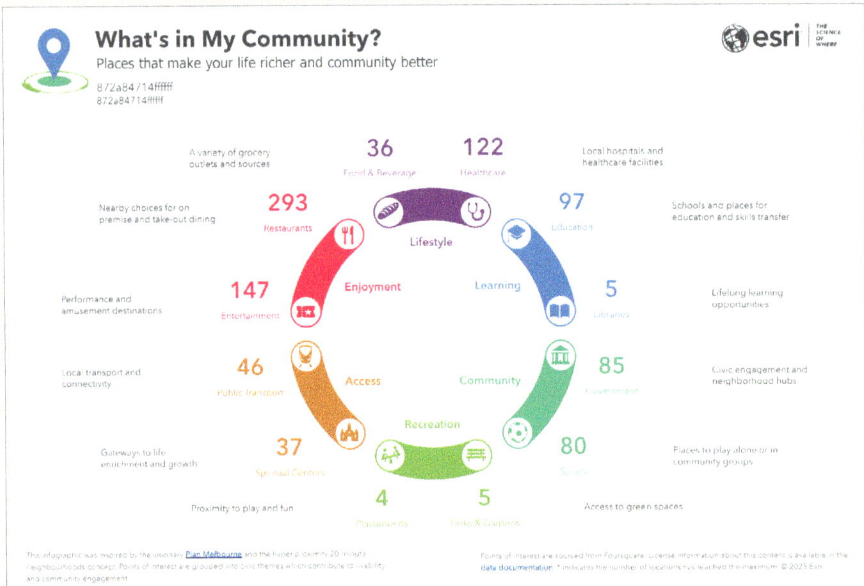


Infographic for the community located in the center of Pittsburgh.

## Summary

In this chapter, you learned how to create a color-coded map and a color-coded layer in ArcGIS Business Analyst. You created a color-coded map of population in Austin, Texas, using Business Analyst Web App. You also created a color-coded layer of population density in Pittsburgh, Pennsylvania, using Business Analyst Pro.

## Workflow

1. Open your project.
2. Set up your map by defining an area of interest and boundary mode/level of detail.
3. Select your variable.
4. Select a color scheme.
5. Explore the results.

# CHAPTER 4

# Running, modifying, and creating infographics

## Coauthor Debashish Ghosh

## Objectives

- Run infographics to compile information on points of interest.
- Modify infographics to match data and themes.

## Introduction

ArcGIS Business Analyst infographics are powerful tools for analyzing and consolidating geographic information and—perhaps just as importantly—for presenting that information to others. Infographics in ArcGIS Business Analyst are interactive, presentation-ready data visualizations.

You can run and create infographics in Business Analyst Web App and Business Analyst Pro. The interface is the same when you're editing an infographic, but the steps to open a project and run an infographic are slightly different in the two applications.

# Tutorial 4-1: Run and create infographics in Business Analyst Web App

In this tutorial, you will use Business Analyst Web App to create infographics about K-12 education in Sedona, Arizona.

> *Important: For Business Analyst Web App, you can run infographics with a Standard license. To modify or create your own infographic, you must have an Advanced license.*

## Create a project

First, you'll create a project.

1. In your browser, navigate to bao.arcgis.com.
2. Enter your ArcGIS credentials and click **Sign In**.
3. On the **Home** tab, click **New project** and name your new project Ch 4 - Infographics.

## Use sites to run infographics

Business Analyst Web App includes Standard and Gallery infographic templates and, with an Advanced license, the ability to design custom templates. Standard infographics are ready-to-use templates. Gallery infographics are a collection of templates designed for specific datasets, communities, and use cases that may not adhere to the general broad appeal of the standard templates. Custom infographics include infographics that you create yourself (My templates) or that are shared with you by your organization (Shared templates).

4. On the map, zoom to Sedona, Arizona.

5. Use the **Pin** tool to create a site in Sedona with default **Rings** radius values.

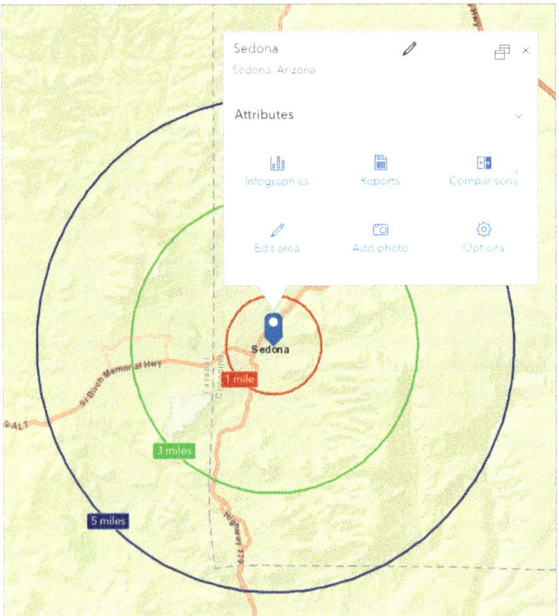


**Tip:** Refer to chapter 2, "Generating Trade Areas," to see how to create sites in Business Analyst Web App.

6. On the pop-up that appears above the pin, click the **Edit** (pencil) button and change the name of the pin to Sedona.
7. At the top of the page, click the **Reports** tab. On the ribbon, click **Run reports** and select **Run infographics**.

   The Infographics workflow appears. Because you have sites open on the Maps tab, they are selected in the Selected sites drop-down list.
8. Click **Add sites**.

> **Tip:** If you have a project with multiple sites, you can select them from the **Recent**, **Currently on map**, **From projects**, and **Neighboring geographies** tabs. You can also click the **Search** (magnifying glass) button to quickly find an existing site. Check the boxes next to the sites that you want to include in the infographic.

9. Make sure that the **Sedona** site is the only one selected. Click **Apply**.
10. Expand the **Standard infographics** and **Gallery infographics** sections and browse the templates they contain.
11. Click the **Search** button and type K-12 Education Overview. In the results, check the box next to the infographic option.
12. Above **Selected infographics**, make sure the **Export infographics** option is turned off.
13. Next to **Sedona**, click **Run now**.

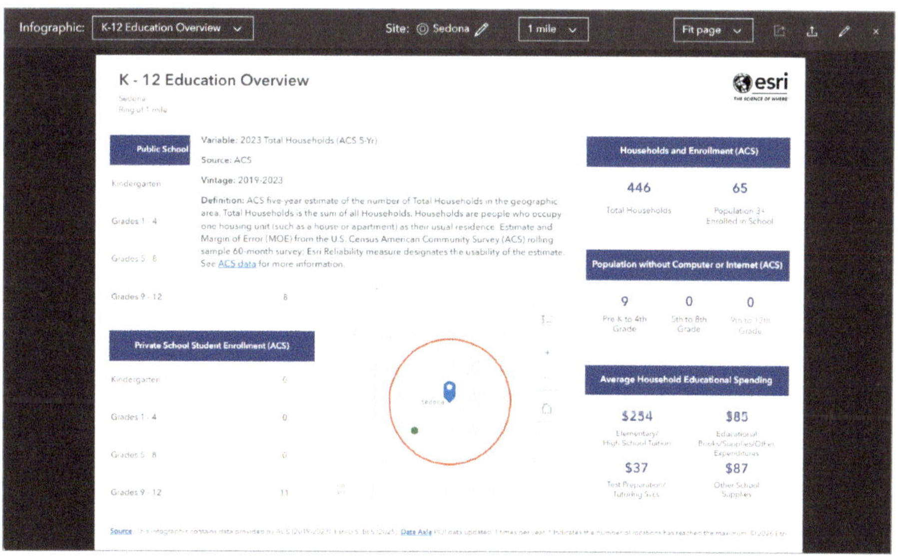


14. At the top of the infographic, click **1 mile** to open the list of ring distances.
15. From the list, click **Compare with geographies**.

    The infographic runs, comparing different geographies to your site, such as comparing Sedona to Coconino County, the state of Arizona, and the United States.

16. Close the infographic.

17. Enable the option **Export infographics**. Make sure that the **Format** is set to **PDF**. Click **Export**.

    You now have a PDF file of the infographic that you configured.

> **Tip:** You can also run an infographic by using your site's pop-up menu. Click on your site to open the pop-up menu and click **Infographics**. Select the infographic of your choice.

## Modify infographics

You can also create your own infographics by modifying an existing template to better suit your needs. When you click Reports and Build infographics, you'll be able to modify Standard, Gallery, and custom templates that you have previously created (My templates) or that were shared with you (Shared templates).

18. On the **Reports** tab, click **Build infographics**.
19. Under **Standard templates**, locate the **Economic Development Profile** template. Hover over the template and click **Open**.
20. Scroll to the second page of the infographic template.
21. Hover over the **Household Income** panel and click **Edit**.
22. In the search field, type monthly mortgage costs. Next to the **2023 HHs w/ Mortgage: Monthly Owner Costs 50+% of HH income (ACS 5-Yr)** variable, select **Percentage**. Drag the variable to the bar chart on the right. Click **Replace the current series**.

23. Repeat the previous step with the following variables as percentages. For each variable, click **Append to current series**.

    - **2023 HHs w/ Mortgage: Monthly Owner Costs 50+% of HH income (ACS 5-Yr)**
    - **2023 HHs w/ Mortgage: Monthly Owner Costs <10% of HH income (ACS 5-Yr)**
    - **2023 HHs w/ Mortgage: Monthly Owner Costs 10-14.9% of HH income (ACS 5-Yr)**
    - **2023 HHs w/ Mortgage: Monthly Owner Costs 15-19.9% of HH income (ACS 5-Yr)**
    - **2023 HHs w/ Mortgage: Monthly Owner Costs 20-24.9% of HH income (ACS 5-Yr)**
    - **2023 HHs w/ Mortgage: Monthly Owner Costs 25-29.9% of HH income (ACS 5-Yr)**
    - **2023 HHs w/ Mortgage: Monthly Owner Costs 30-34.9% of HH income (ACS 5-Yr)**
    - **2023 HHs w/ Mortgage: Monthly Owner Costs 35-39.9% of HH income (ACS 5-Yr)**
    - **2023 HHs w/ Mortgage: Monthly Owner Costs 40-40.9% of HH income (ACS 5-Yr)**

24. Click the title of the chart. Replace **Title** text with Monthly Mortgage Costs.

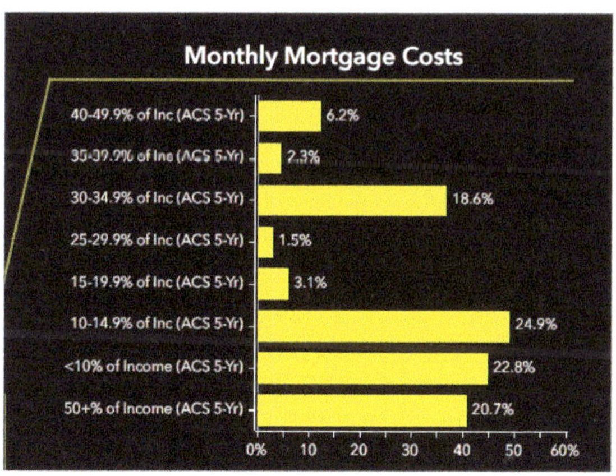


25. Click **Apply**. Save the infographic.

    The template is now updated with a new chart.

# Tutorial 4-2: Run and create infographics in Business Analyst Pro

In this tutorial, you'll create an infographic about soda shops in Salt Lake City, Utah, using Business Analyst Pro, which has a different setup than in Business Analyst Web App.

## Create a project

1. Start ArcGIS Pro. If prompted, sign in using your licensed ArcGIS organizational account.
2. Under **New Project**, click **Map**. In the **New Project** dialog box, apply the following settings:
   - **Name**: Ch 4 - Infographics
   - **Location**: C:\Top20BusinessAnalyst

   **Tip:** This book uses **C:\Top20BusinessAnalyst** for the folder location, but you can use any folder.

3. Click **OK**.

   The project is created with a default map.

## Perform a points of interest search

4. Zoom to the general area of Salt Lake City, Utah.

5. On the ribbon, click the **Analysis** tab.

6. In the **Workflows** group, click **Business Analysis** to open the list of tools. Under **General Workflows**, select **Points of Interest Search**.

   The Points of Interest Search pane appears.

7. For **Step One: Initial Search**, under **Name or keyword**, type soda and click **Next**.

8. For **Step Two: Filter**, beneath the results list, click **Preview Map** to show all the soda shops the search returned.

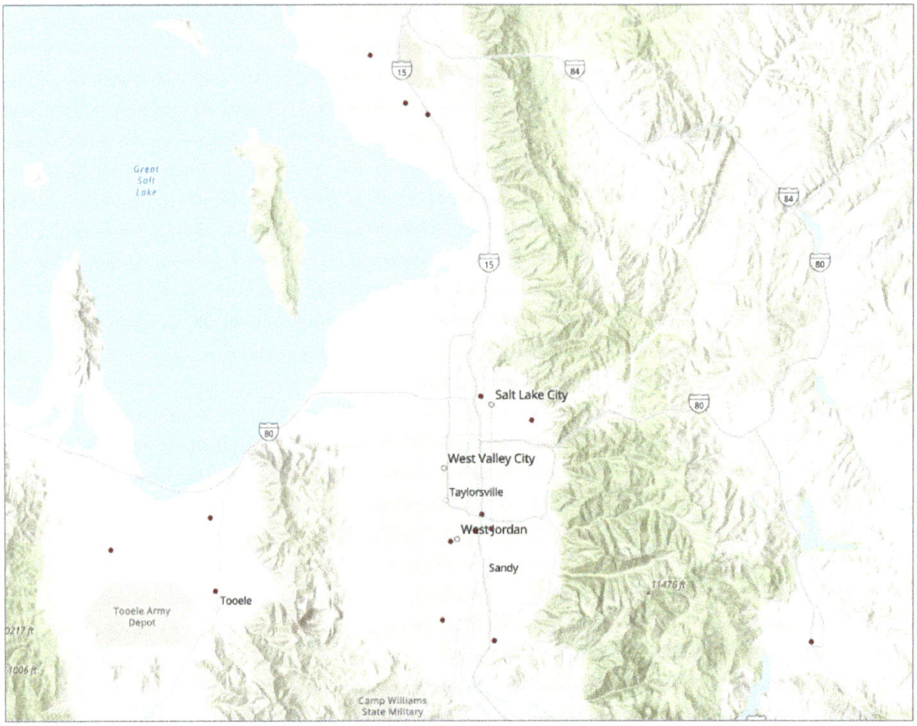


9. In the results list, expand **City**. Hover next to **Salt Lake City** and click **Only**.

10. For **Output**, change the layer name to Soda_Salt_Lake_City.

11. Click **Finish**.

The soda shop locations are saved to a new layer, visible on the map and in the Contents pane.

## Generate trade areas for soda shops

With the soda shop locations on the map, you will create drive time trade areas around each one.

12. On the ribbon, click the **Analysis** tab. In the **Geoprocessing** group, click **Tools**.
13. In the **Geoprocessing** pane, search for Generate Drive Time Trade Areas and click the tool to open it.
14. In the **Generate Drive Time Trade Areas** tool, apply the following settings:
    - For **Input Features**, select **Soda_Salt_Lake_City**.
    - For **Output Feature Class**, change the name to Soda_Salt_Lake_City_drivetimes.
    - For **Distance Type**, leave the default, **Driving Time**.
    - For **Input Method**, leave the default, **Values**.
    - For **Distances**, type 5.
    - For **Distance Units**, leave the default, **Minutes**.
    - For **ID Field**, select **Esri PID**.
    - For **Dissolve Option**, leave the default, **Overlap**.
    - At the top of the tool, click **estimate credits**.

    The tool is estimated to consume about 1.5 credits.

15. Click **Run**.

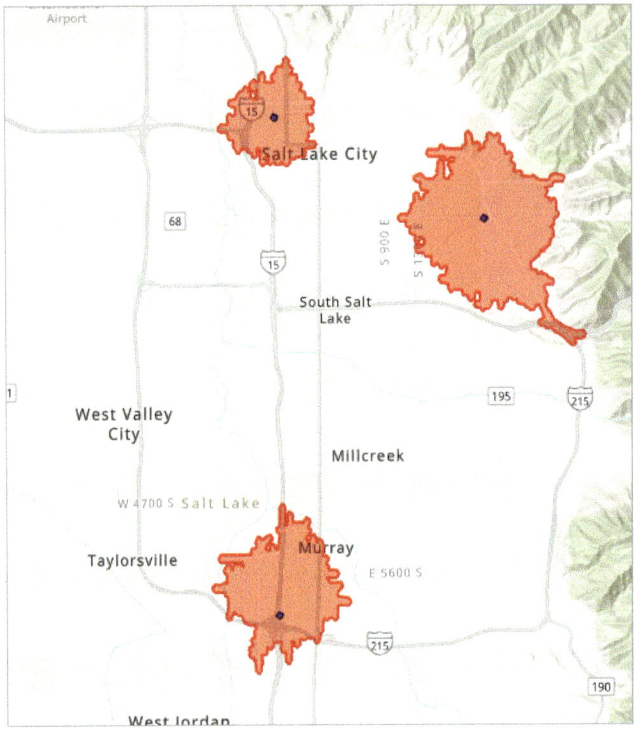


The drive-time trade areas are added to the map.

## Use the Summary Reports tool to run infographics

16. In the **Geoprocessing** pane, search for and open the **Summary Reports** tool.

17. In the **Summary Reports** tool, apply the following settings:

- For **Boundary Layer**, select **Soda_Salt_Lake_City_drivetimes**.
- For **Create Reports**, type Demographic Profile.
- For **Output Folder**, type Soda_Salt_Lake_City_Demographic_Profile
- At the top of the tool, click **estimate credits**.

The tool is estimated to consume 10 credits.

18. Click **Run**.

The selected infographics are run and saved to the chosen output folder.

19. At the bottom of the **Geoprocessing** pane, click **View Details**. Click the **Parameters** tab. Next to **Output Folder**, click the link to open the folder where the infographic is saved.

20. In the folder, double-click the **Demographic Profile** PDF file.

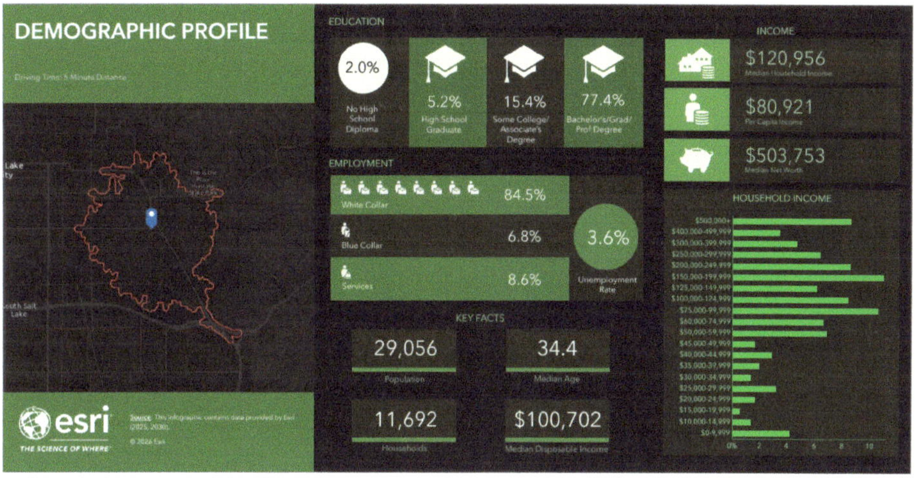


The Demographic Profile has a page for each soda shop in Salt Lake City. It contains demographics about the surrounding area, such as education, employment, and income information.

**Tip:** You can also run infographics from the **Map** tab, in the **Inquiry** group, by clicking **Infographics** to interactively view infographic templates based on the selection on your map.

## Take it to the next level

There are many best practices to consider when creating your own infographics, especially for text, theming, and charts and tables.

### Best practices for text

The Business Analyst infographic builder provides a curated selection of fonts that are legible and popular, while still offering room to add your own flavor to the design. Learn more at link.esri.com/Top20BA/Text.

### Font

It's recommended to minimize the number of fonts in any one template. Even if you're going for a quirky feel or trying to differentiate spaces on the template, too many fonts can prove distracting and jumbled. The best practice is to stick to no more than three font types or styles in a template.

### Text size

You may want to vary text size based on the meaning or importance of the text. You can create a hierarchy of panel headers, text blocks, and captions for maps or images. The ArcGIS Business Analyst team uses the following guidelines:

- Header text: 26-point minimum
- Body text: 14-point minimum
- Caption text: 12-point minimum

### Font color

As with text size, be purposeful about the number of colors you use for text in infographics. The best practice is to use up to three complementary colors in an established hierarchy (primary, secondary, tertiary) to communicate importance.

### Title

A well-formatted title can go a long way in conveying polish and providing viewers of your infographic with the information they need. If you're looking for guidance on title sizing, here are some suggestions:

- Infographic title: 36-point minimum
- Site name: 16-point minimum
- Site description: 14-point minimum

### Dynamic fields

We recommend that you create a title panel that includes not just the title, but also dynamic fields that populate with information, such as the site name (the location where you are running the infographic) and the site description (the size and type of site, such as a 20-minute drive time).

### Logo

You can add your logo—even multiple versions of your logo—in the infographic preferences. Be sure to use a logo that matches the color theme of your template and size it for the best legibility. The best practice is a 28-pixel minimum height.

### Footers

Footer text is essential for providing thorough, accurate data attribution. We recommend using the Data source description option, which includes the data source and vintage for all data in the template. When you create a footer for your infographic, the suggested font size is 12 point for legibility.

## Best practices for tables and charts

Tables make great use of real estate in an infographic: You can display a large quantity of data while keeping the format neat, organized, and legible. Learn more at link.esri.com/Top20BA/Tables.

There are a few types of tables you can add to infographics:

- **Tables with variables:** Select variables in the data browser and display the data for your sites.
- **Comparison tables:** Select variables in the data browser and display the data for your sites, as well as for geographies that you choose.
- **Nearby tables:** Select a location type (such as businesses or other points of interest) and display the locations near your site.
- **Custom tables:** If you have your own requirements for a table, you can create a custom table to fit your specifications.

### Size elements correctly

When it comes to making useful tables, the best advice is also the simplest: Make them easy to read. It can be tempting to reduce the size of the text so you can squeeze tons of data into a table. The best practice is the following:

- **Font size:** 12-point minimum
- **Row height:** 24-point minimum

### Justify columns

There are no set rules about text justification, but the ArcGIS Business Analyst team at Esri often defers to the following guidelines:

- The variable name (usually listed in the left-most column) justified left.
- Numeric data (indicating variable values) justified right.

### Row spacing

Use cell styling to increase the bottom border of a row to a larger thickness to increase spacing between rows.

### Tabular infographic styling

Tabular infographic styling supports theming and modern design options. Tabular infographics are designed to display larger font sizes and high color contrast to meet accessibility standards. This improves readability for a variety of audiences. The default color scheme is designed to be print-friendly.

### Choose the right type of chart

The first thing to consider is the size of the panel within the template. If you've got plenty of room to work with, having a big bar chart might be the right call; if you've only got a tiny space open, a pie chart could be a better fit. Sometimes letting the shape of the space guide you is best.

You'll also want to attend to the number of variables and series needed. For example, pie charts should only be used with up to five variables—and those variables should be a complete collection because "pies" are generally interpreted as comprising 100%.

### Label your axes

When visualizing data in bar charts or line charts, be sure that all axis labels are legible.

## Best practices for series

Organizing your chart data into series is a way to make grouped information clear and legible. Series are used for showing information about subgroups of main categories. A separate column or line represents each of the subgroups, which are displayed in different colors to distinguish them.

## Best practices for themes

The theme of an infographic template consists of colors, fonts, and background image settings. It gives the template a unified look across its multiple panels and pages. You can also use the theme to give a template an appearance that helps visually tie it together with your organization's website or publicity materials. Learn more at link.esri.com/Top20BA/Themes.

### Color

Use color to communicate meaning and emphasis rather than decoration. Use color consistently to create a familiar and engaging experience. For greater impact, use color sparsely.

### Branding

When the use of your organization's brand color is desired:

- Apply it to emphasized and prioritized elements.
- Avoid overusing it or using it as a background.

### Text

Select a text color that is legible when displayed over the selected background color. Exercise care when selecting a background image that differs from the background color. Use Contrast Checker (link.esri.com/Top20BA/Contrast) to verify that your choice of colors meets accessibility requirements. You can also consider testing the contrast of the template using appropriate tools (for example, developer tools in the Chrome browser) to emulate vision deficiencies.

## Summary

In this chapter, you ran and explored infographics in Business Analyst Web App and Business Analyst Pro. You also modified an existing infographic in Business Analyst Web App. You  learned the best practices for designing your own infographic.

### Workflow

1. Open your project and add your sites.
2. Run and explore infographics.
3. Modify infographics to match your data and theme.

# CHAPTER 5

# Mapping points of interest

Coauthor Yann Bovet

## Objectives

- Search for points of interest (POIs).
- Filter your results to find specific POIs.
- Explore the area surrounding POIs.

## Introduction

You can search for points of interest, such as businesses and facilities, in ArcGIS Business Analyst. For example, for health care, you can search for hospitals or search using the name of a specific hospital. You can also search within a category using an industry list or search using North American Industry Classification System (NAICS) codes.

# Tutorial 5-1: Use a POI search in Business Analyst Web App

In this tutorial, you will use a POI search to analyze parking and entertainment in Lausanne, Switzerland. From there, you can identify where places for entertainment are located and whether there is ample public parking nearby.

## Create a project

1. In your browser, navigate to bao.arcgis.com.
2. Enter your ArcGIS credentials and click **Sign In**.
3. On the **Home** tab, click **New project**. Name your new project Ch. 5 - Map POIs and click **Create**. Open your project.

   **Tip:** If prompted to clear the map of previous data, click **Yes**.

## Find points of interest using the POI search workflow

In Business Analyst Web App, you can set your geography; set your POI data source; search for POIs using keywords, categories, or codes; prefilter your search; filter your search results; and explore the results.

4. At the top of the page, click **Data source** to open the list of sources and select **Switzerland**.
5. Click the **Maps** tab. On the **Maps** tab, click **Create maps** and click **Points of interest (POI) search**.
6. In the **Points of interest (POI)** search pane, expand the **Geography** section. Click the **Area of interest** field and search for Vaud.
7. In the results, select **22 (Vaud) Canton**.

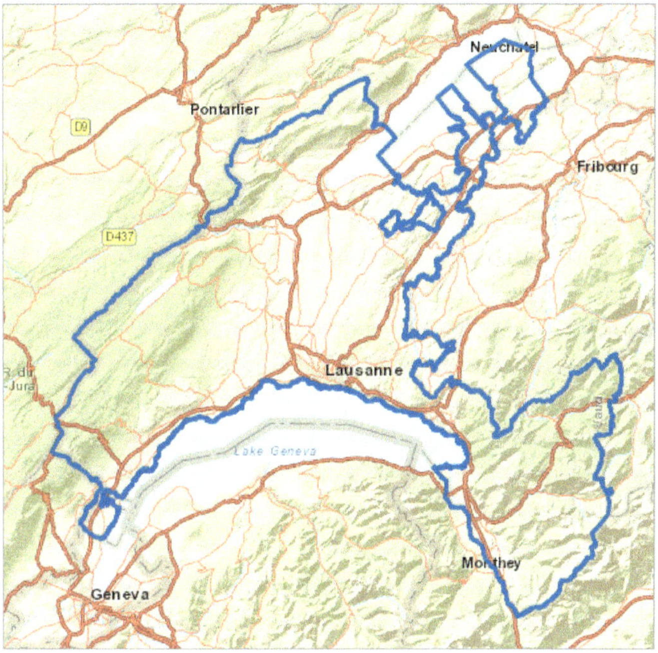


8. Click the **Search** field and click the **Advanced search** tab.

You are interested in searching for all places for entertainment and parking in the area. To add a condition, you must select a field, choose a filter operator, and set a value.

9. Click **Add condition**. Create a condition as follows:

- For the field, click **Foursquare Category**.
- For the filter operator, make sure **is any of the following** is selected.
- For the value, check the box for **Arts and Entertainment (All)**. In the **Enter category name** field, type parking and check the box for **Parking**.

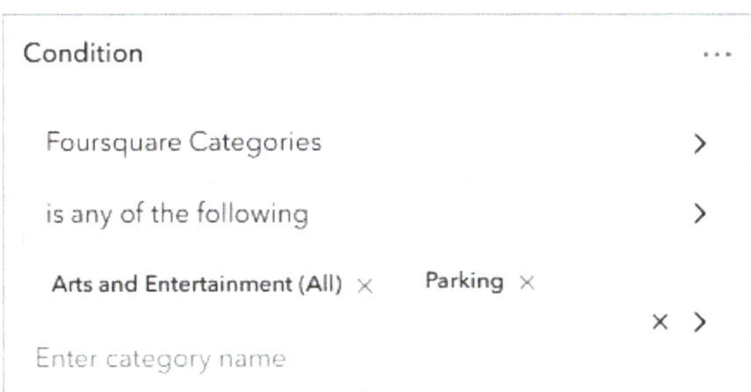

10. Click **Search**.

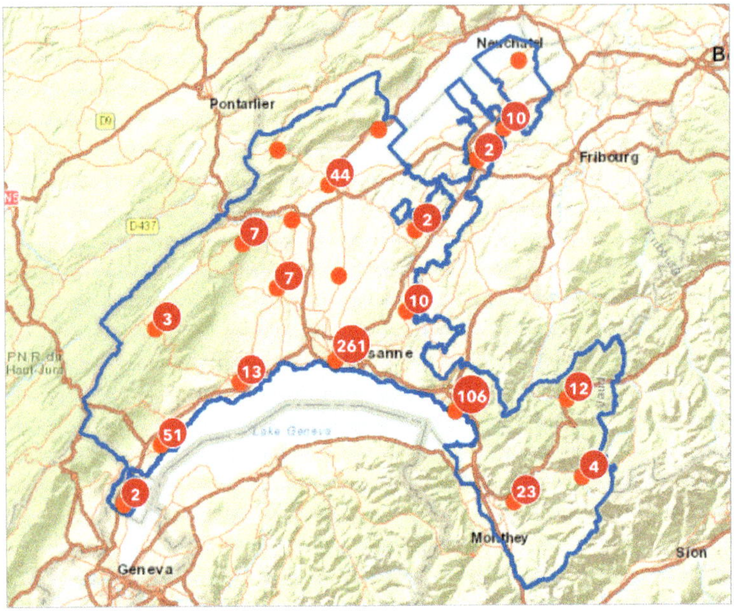


The search returns approximately 544 point locations. You can also use another advanced search condition to make your search more complex and your results more specific. Because there is a restriction on returning a maximum of 5,000 points when mapping with an online data source, using an advanced search condition is useful to make the query as specific as possible for the most relevant searching.

11. Click the **Search** field. Click **Add another condition**.
12. Create a condition as follows:
    - For the field, click **Locality**.
    - For the filter operator, click **is**.
    - For the value, type Lausanne.
13. Click **Search** to perform the POI search with the new advanced condition.
14. Zoom in to explore the results.

You can also use a filter to further refine your results. Although prefiltering uses conditions to create a more specific search, filtering refines the search results to return more specific places. Now check the options in postcodes 1003, 1004, and 1005.

15. In the **Points of interest (POI) search** pane, in the **Filters** section, click **Show filters**.

16. Expand the **Postcode** attribute. Check the boxes for postcodes **1003**, **1004**, and **1005**.

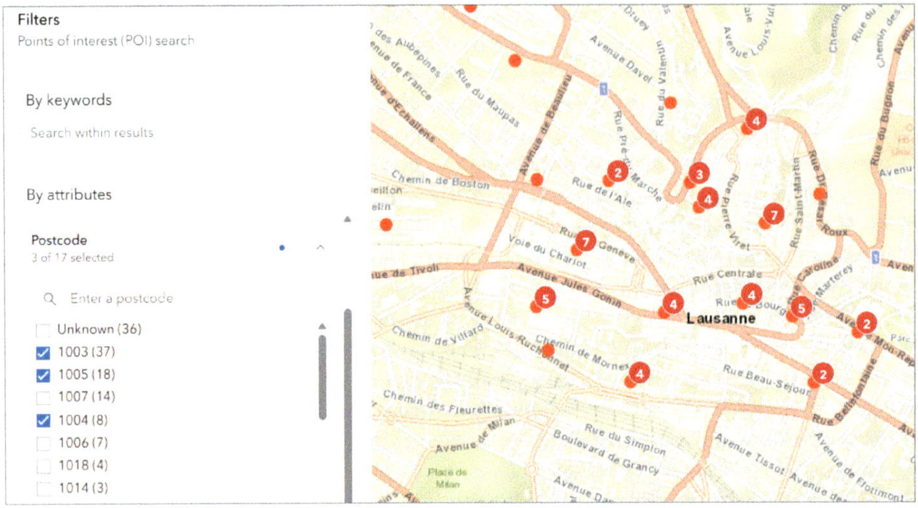


The filter is automatically applied, and the map results are updated.

17. Above **Filters**, click the back button.

18. In the **Style** section, change the style to **Place** if it is not already set.

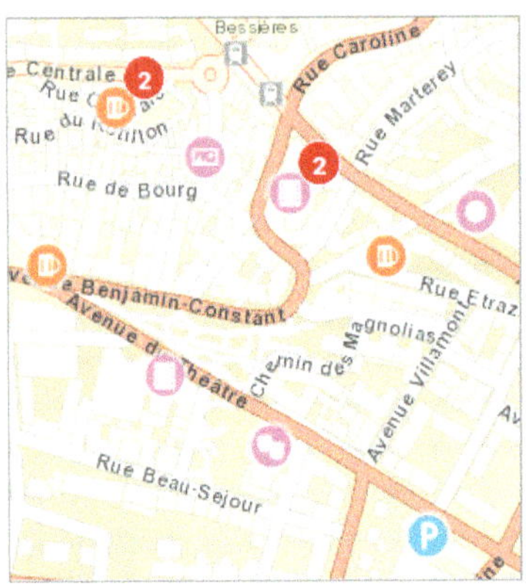


The point icons are replaced with curated icons that match the location types. Now you'll increase the icon sizes and cluster the points to aggregate them.

19. Click **Show style options**. For **Size**, type 35. Click the back button.

> **Tip:** The **Cluster points** option allows you to aggregate points into clusters or display them individually on the map.

20. In the **Legend** section, find and click the icon for parking (blue circle with a *P*).

21. In the **Select symbol** window, for **Size**, drag the slider to increase the parking icon size to **50**. Click **Done**.

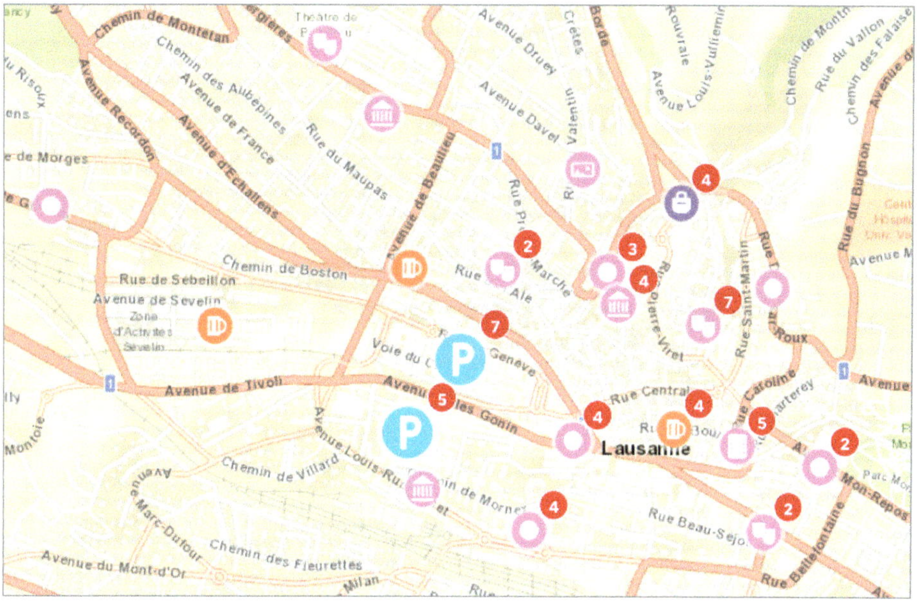


Now all the parking icons are the largest icons on the map. Use the map legend to reference what the icons symbolize and how many places are identified on the map.

Next, you'll explore the Results pane.

22. On the **Summary** tab, use the horizontal scroll bar to see the top five primary industries.

23. Click the **Table** tab to view the results as a table. Hover over an item in the table to highlight the corresponding site on the map.

> **Tip:** You can use the **Customize table** option to modify the columns in the table. Drag the name of a column to change its order in the table. Uncheck the box next to the column name to deselect it from the table.

### Take it to the next level

Points of interest can be used alone or in tandem with additional data to enrich the analysis. For instance, run infographics for your sites.

24. When you have finished exploring the POIs, change the **Data Source** back to your original source (for example, **USA Esri 2025**).

## Tutorial 5-2: Use a POI search in Business Analyst Pro

In this tutorial, you will map electric vehicle charging stations around Los Angeles, California, using the POI search workflow in Business Analyst Pro to identify patterns in density.

### Create a project

You will create a new project in Business Analyst Pro.

1. Start ArcGIS Pro. If prompted, sign in using your licensed ArcGIS organizational account.
2. Under **New Project**, click **Map**.
3. In the **New Project** dialog box, apply the following settings:
    - **Name**: Ch 5 - Map POIs
    - **Location**: C:\Top20BusinessAnalyst

    **Tip:** This book uses **C:\Top20BusinessAnalyst** for the folder location, but you can use any folder.

4. Click **OK**.

    The project is created with a default map.

## Map points of interest

In Business Analyst Pro, you can set your data source; search using a search term, categories, or codes; refine your results using filtering; and visualize the results on the map and in a table.

5. Zoom to the general area of Los Angeles, California.

6. On the ribbon, click the **Analysis** tab.
7. In the **Workflows** group, click **Business Analysis** to open the list of workflows and click **Points of Interest Search**.

   The Points of Interest Search pane appears.
8. For **Step One: Initial Search**, under **Name or keyword**, type charging station and click **Next**.

9. For **Step Two: Filter**, under **Company/Business Name**, hover next to **Electric Charging Station** and click **Only** so that it is the only result included.

10. Expand the **City** section. Next to **Los Angeles**, click **Only**.

11. For **Output**, change the layer name to charging_station_LA. Click **Finish**.

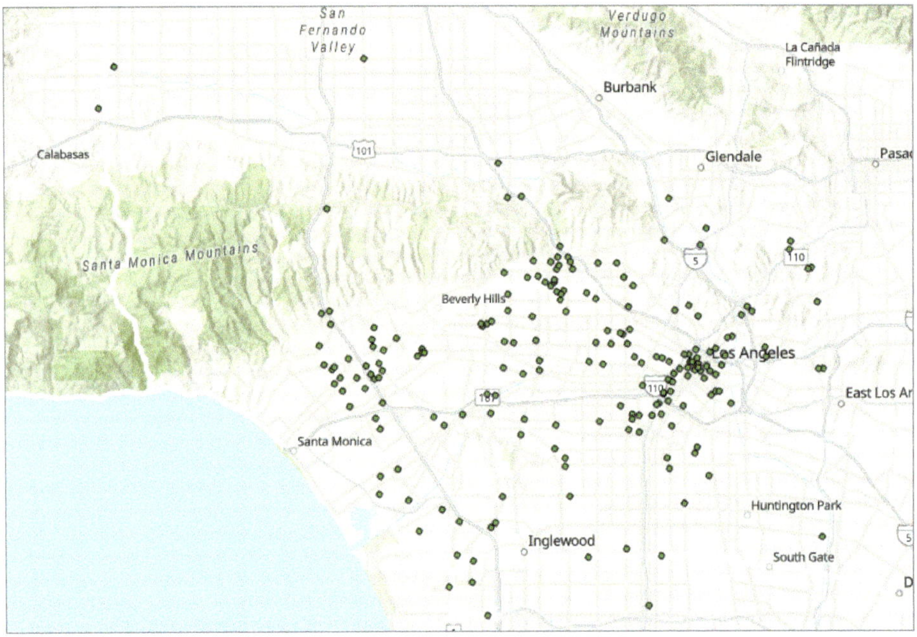


The results appear on the map. Each dot represents a charging station in Los Angeles. You can use this information to visually identify patterns, such as whether certain areas have a higher or lower density of electric charging stations to determine where you can find a competitive location for your new charging station.

## Summary

In this chapter, you mapped parking and entertainment POIs in Switzerland using Business Analyst Web App. Using Business Analyst Pro, you mapped EV charging stations around Los Angeles, California. Although the essential skill is shared between both applications, the steps are slightly different.

### Workflow

1. Open your project and set your data source.
2. Select the area of interest.
3. Search for POIs.
4. Use conditions and filters to refine your search.
5. Explore the results.

# CHAPTER 6

# Using data

## Coauthor Leann Kurias

## Objectives

- Use data to visualize information about a site.
- Work with point location data in an added layer.
- Set up custom data to run benchmark comparisons.

## Introduction

In this chapter, you will learn how to add data to the map in ArcGIS Business Analyst. First, you will add a layer of point locations in Business Analyst Web App. After converting selected locations into sites, you will create a color-coded map illustrating demographic information in these areas. You will then set up custom data from a federal agency in Business Analyst Pro and use a benchmark comparison to gain a broad view of the data.

# Tutorial 6-1: Add and work with data in Business Analyst Web App

In this tutorial, you will add data using World Transit Stop data and explore stops in Flagstaff, Arizona, using Business Analyst Web App.

## Create a project

1. In your browser, navigate to bao.arcgis.com.
2. Enter your ArcGIS credentials and click **Sign In**.
3. On the **Home** tab, click **New project**. Name your project Ch. 6 - Public transport data and click **Create**. Open your project.

**Tip:** If you are prompted to clear the map of previous data, click **Yes**.

## Browse and add data

You will search for and add a layer of transit stops in ArcGIS Living Atlas of the World.

4. Make sure that the current **Data Source** is set to **USA (Esri 2025)**.
5. On the **Maps** tab, click **Add data**. Click **Web maps and layers**.
6. In the **Web maps and layers** window, under **Explore,** expand the **Living Atlas** section. Click **Show all.**
7. Below the search bar, click the **Regions** drop-down menu. Check the box for **United States** and click **Done**.
8. Click the search bar and type transit.

9. In the search results, hover over the **World Transit Stops** layer and click **Add to map**.

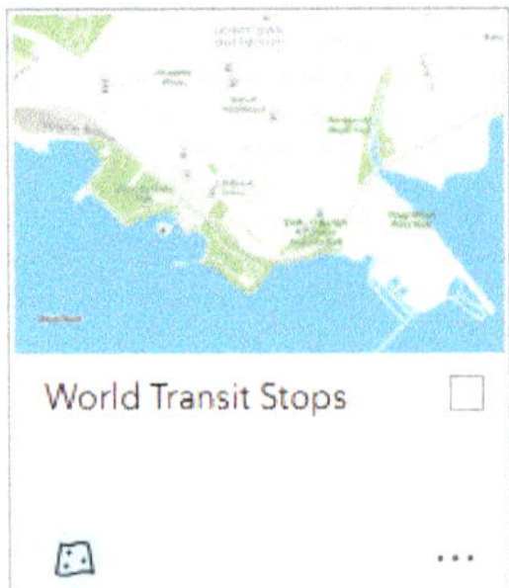


10. On the **Maps** tab, locate the search bar and type Flagstaff. Select **Flagstaff, AZ, USA**.

11. Zoom in until you can see the data layer.

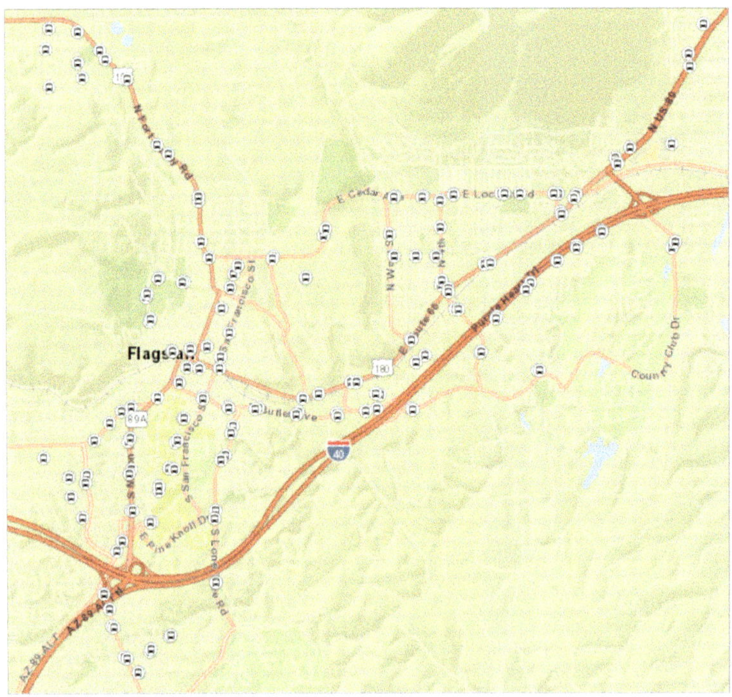

12. On the right, in the map tools, click the **Select** tool.

13. On the map, click a transit stop. Press and hold the Shift key and click another stop some distance away to select them.

14. In the pop-up menu of the second transit stop, click **Create buffers**.

15. Click the **Walk time** tab. For **Time**, remove the second and third values so that there is only the first value, **5**. Click **Apply**.

16. In the pop-up menu, click **Create sites** and then click **Create 2 sites**.

17. In the **Save created sites** window, select **Create a new layer**. Name the layer Flagstaff Transit Stop Locations and click **Save**.

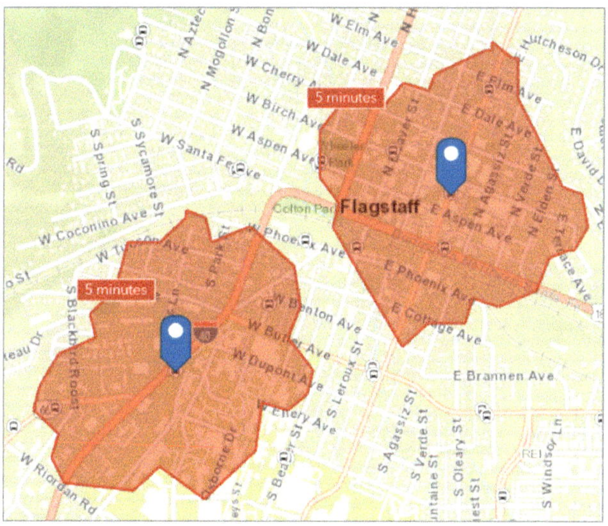


## Explore the data

To learn more about the locations you mapped, you will create a color-coded map.

18. On the **Maps** tab, click **Create maps** and click **Color-coded maps**.

19. In the **Color-coded maps** pane, expand the **Geography** section and click the **Area of interest** field.

20. Click **Your sites** and then select the two transit stops you created sites for. Click **Apply**.

21. For **Level of detail**, leave **Automatic (Block Groups)** selected.

22. At the bottom of the pane, click **Browse all variables**.

23. In the search bar, type Workers 16+ Took Public Transportation and click **Search**.

24. Click the **2023 Workers 16+ Took Public Transportation (ACS 5-Yr)** variable and leave **Count** as the calculation type.

**Tip:** If the 2023 variable is not available, select the most recently published variable.

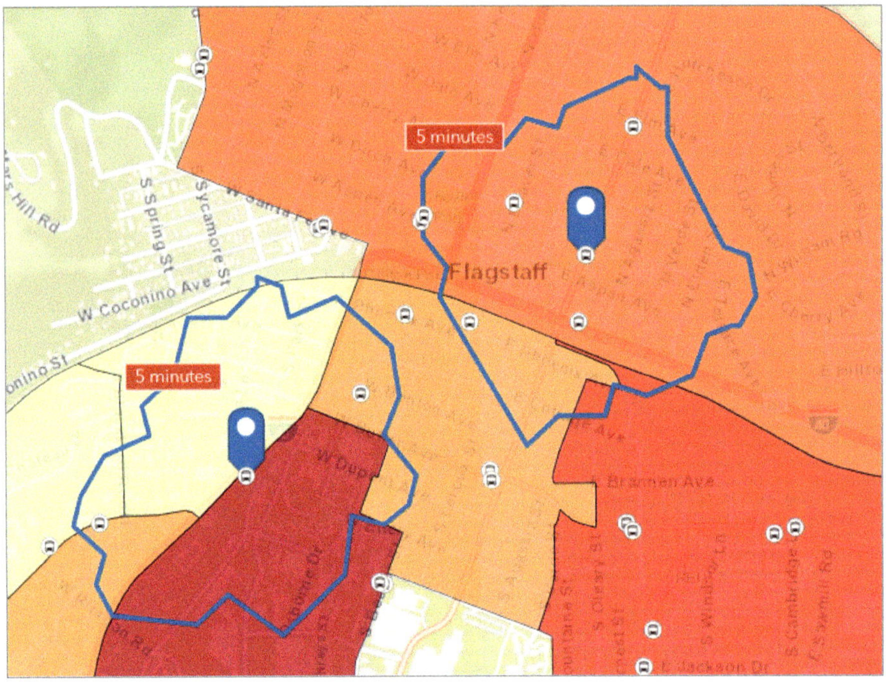


The map now shows block groups that intersect with the five-minute walk times near the transit stops selected, color-coded by the number of potential transit riders.

## Take it to the next level

Now that you have learned about public transit ridership near these two transit stops, you can learn more about the commuting habits of these block groups by running infographics. Click one of the sites and click **Infographics**. In the **Infographic** drop-down menu, click **Commute Profile**.

The infographic provides an overview of types of transportation taken to work as well as the number of workers and average travel times to work.

# Tutorial 6-2: Set up custom data in Business Analyst Pro

In this tutorial, you will set up custom data using Federal Emergency Management Agency (FEMA) data in Business Analyst Pro.

> *Important: To complete this tutorial, you must connect to a locally stored ArcGIS Business Analyst dataset.*

## Create a project

1. Start ArcGIS Pro. If prompted, sign in using your licensed ArcGIS organizational account.
2. Under **New Project**, click **Map**.
3. In the **New Project** dialog box, apply the following settings:
   - **Name**: Ch 6 - Using custom data
   - **Location**: C:\Top20BusinessAnalyst

   **Tip:** This book uses **C:\Top20BusinessAnalyst** for the folder location, but you can use any folder.

## Download the data

You will download risk data from an OpenFEMA dataset.

4. Go to the FEMA National Risk Index site at link.esri.com/Top20BA/FEMA.
5. Scroll down to the **Available Downloadable National Risk Index Data** section.

6. From the table, download the **Geodatabase** version of the **All Counties** dataset.

| Dataset | Table Format | Shapefile | Geodatabase |
|---|---|---|---|
| All States and Territories | Download | Download | Download |
| All Counties | Download | Download | Download |
| All Census tracts | Download | Download | Download |
| Tribal County Relational Database | Download | Download | Download |
| Tribal Census tract Relational Database | Download | Download | Download |

7. In File Explorer, go to the **Downloads** folder and locate the **NRI_GDB_Counties.zip** file. Extract its contents to **C:\Top20BusinessAnalyst\Ch 6 - Using custom data**.

8. Once extracted, browse to the folder you selected, open the **NRIDataDictionary.csv** table, and review it.

   The spreadsheet explains the field name and the data type for each field of the data.

9. Close the table when you have finished.

   The National Risk Index (NRI) dataset comprehensively assesses various hazards and their associated risks across different geographic areas. Key fields in the dataset include Federal Information Processing Standards (FIPS) codes for geographic identification, overall risk ratings and scores, social vulnerability, resilience scores reflecting community capacity, annual losses, and risk scores for various hazards. The NRI covers a range of hazards, such as flooding, earthquakes, wildfires, severe storms, landslides, and droughts.

   Each entry helps illustrate how communities might be impacted by various natural disasters, enabling better preparedness and response strategies. These variables are later categorized to improve use for analysis.

## Set up the data

To use this data in Business Analyst Pro, you will create a statistical data collection.

10. On the ribbon, click the **Analysis** tab. In the **Workflows** group, click **Business Analysis**.

11. At the bottom of the menu, under **Business Analyst Data Source**, click **Change data source**. Set to a local dataset, such as **USA_ESRI_2025**.

12. In the **Business Analysis** tools, under **Statistical Data Collections**, click **New Statistical Data Collection**.

13. In the **Create Statistical Data Collection** window, for **Input Data**, click the browse button (yellow folder icon).

14. Browse to where you stored the **NRI_GDB_Counties.gdb** file and double-click it. Double-click **NRI_Features** and click **NRI_Counties**. Click **OK**.

15. Leave the **Output Statistical Data Collection** name as the default, **NRI_Counties.sdcx**, and click **Create**. If it appears, accept the GEOM warning and click **OK**.

    The NRI_Counties window appears. Here, you can explore the data's variables and properties and modify properties. Next, you'll create a performance index.

16. Scroll down to the **Performance Index** section. Click **Build Index** and then click **OK**.

    After a few minutes, the performance index is built.

17. In the **NRI_Counties** window, click **OK**.

18. In the **Catalog** pane, expand **Business Analyst** > **Statistical Data Collections** to view the new collection.

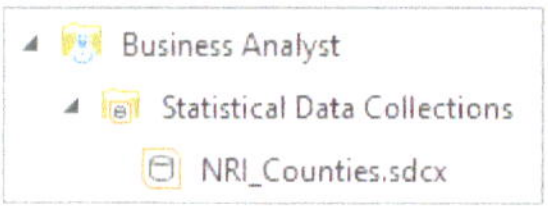


## Run benchmark comparisons

You will use a benchmark comparison to explore risk data in California at the county level.

19. On the **Analysis** tab, in the **Geoprocessing** group, click **Tools**.

20. In the **Geoprocessing** pane search bar, type Generate Standard Geography Trade Areas. In the list of results, click **Generate Standard Geography Trade Areas**.

21. In the **Generate Standard Geography Trade Areas** tool, apply the following settings:

- For **Geography Level**, select **Counties (US.Counties)**.
- For **Output Feature Class**, type California_counties.
- For **Geography IDs List**, click the browse button. In the **Select Geography: US.Counties** window, select **US by Counties**. Select **06 California**. Check the **06 California** box to select all 58 counties. Click **OK**.

22. Click **Run**.

The feature layer of California counties is added to the map and the Contents pane.

23. On the **Analysis** tab, click **Business Analysis** to open the gallery and click **Benchmark Comparisons**.

24. In the **Benchmark Comparisons Properties** tool, apply the workflow settings as follows:

- For **Input Features**, select **California_counties**.
- For **Name Field**, select **NAMES**.
- Next to **Variables**, click the plus button.

25. In the **Data Browser** window, click the **Custom Data** tab and click **NRI_Counties**. From the list, select the following variables:

- **National Risk Index - Score - Composite**
- **Social Vulnerability - Score**
- **Community Resilience - Score**
- **Coastal Flooding - Hazard Type Risk Index Score**
- **Drought - Hazard Type Risk Index Score**
- **Heat Wave - Hazard Type Risk Index Score**
- **Wildfire - Hazard Type Risk Index Score**

26. In the upper right of the window, click the **Show/Hide details panel** button and click **Save List**. For **Name**, type Risk Assessment Variables and click **OK**.

27. In the **Data Browser** window, click **OK**.

The list of variables appears in the Benchmark Comparisons Properties pane.

28. At the top of the tool pane, click the **Comparison Methods** tab. Apply the workflow settings as follows:

- For **Statistics**, check the box for **Mean**.
- For **Method**, select **Above and below benchmark**.
- For **Benchmark**, select **Mean**.

29. Click **Apply**.

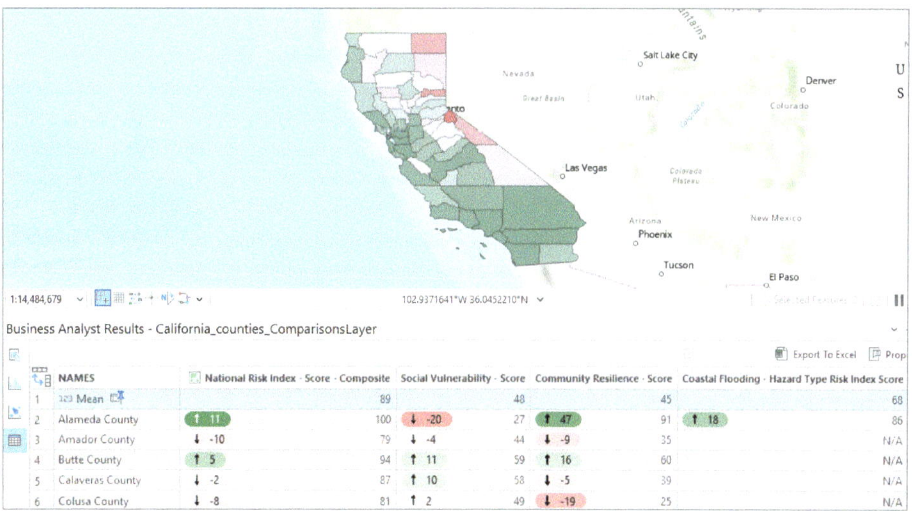

The custom variables are applied to the benchmark comparison. The results are displayed as a table in the Results pane. The results and the map automatically update to show how each county's risk compares with the mean. Counties in green are above the statewide mean for the composite risk score, and counties in red are below this benchmark. You'll reverse the color scheme so that counties above the statewide mean display as higher risk.

30. In the **Benchmark Comparisons Properties** pane, for **Style**, click the **Color scheme** list and click **Color scheme properties**.

31. In the **Color Scheme Editor** window, click the **Reverse color scheme** button.

32. Click **OK** and click **Apply**.

    The colors are now more intuitive, indicating higher risk with red polygon symbolization.

33. In the upper left of the **Results** table, click the **Transpose Table** button to switch the table's rows and columns.

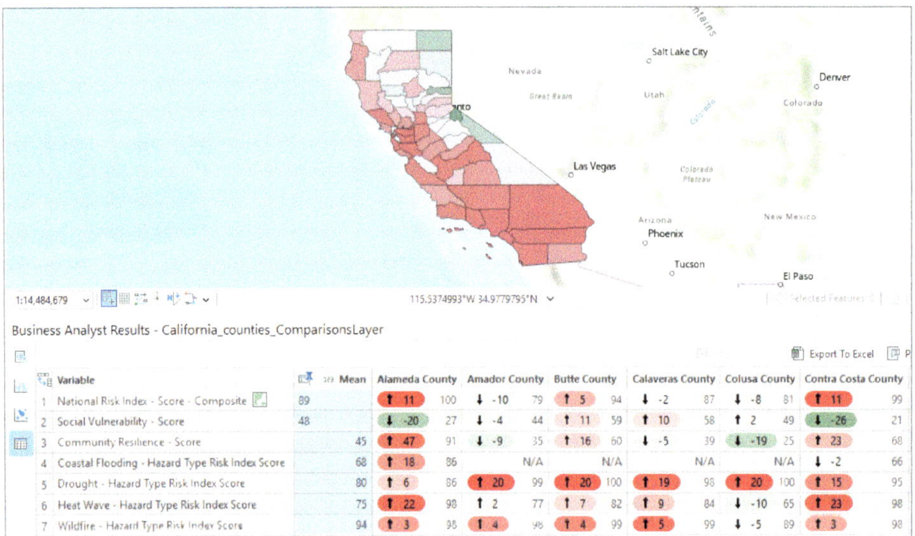


Counties with higher or lower risk are color-coded, offering a clear view of the disparities in the table. You can view this information in ArcGIS Pro or export the results to Microsoft Excel.

34. Close the **Results** pane and save your project.

### Take it to the next level

You can also use custom data in infographics. Infographic panels serve as individual sections within the larger infographic, allowing you to present specific data points clearly and effectively. For example, you can add an infographic panel with the custom data. By choosing to add panels, you can bring focus to various aspects of the housing market that you want to highlight, such as vulnerability to climate risks or changes in property values.

You might add an infographic panel to the **Housing Market Characteristics** template that provides an overview of a site's risk outlook, highlighting custom variables you set up from the NRI data.

## Summary

In this chapter, you learned how to add data to a map in ArcGIS Business Analyst. You added a layer of point locations in Business Analyst Web App, converted locations into sites, and created a color-coded map. You also set up custom data from a federal agency in Business Analyst Pro and used it in a benchmark comparison.

### Workflow

1. Add data to the map by specifying a data source, such as an online layer or file geodatabase.
2. Configure data for use in ArcGIS Business Analyst.
3. Use data in ArcGIS Business Analyst workflows, such as color-coded maps and layers or benchmark comparisons.

# CHAPTER 7

# Sharing your work

## Objectives

- Create a dashboard to share your ArcGIS Business Analyst infographic.
- Create a story to share your ArcGIS Business Analyst sites, maps, and infographics.

## Introduction

In Business Analyst Web App Advanced, you can create dashboards and stories to share infographics from ArcGIS Business Analyst in different ArcGIS formats.

A dashboard is a view of geographic information that displays multiple visualizations on one screen, allowing you to provide a comprehensive view of data to support rapid decision-making. You can use infographics in Business Analyst Web App to visualize key indicators for sites in the form of charts, graphics, tables, images, and text. The ArcGIS Dashboards workflow allows you to choose a dashboard type and layout and select sites and infographic templates to create a dashboard from Business Analyst Web App.

ArcGIS StoryMaps is a web app for storytelling. Using the ArcGIS StoryMaps workflow in ArcGIS Business Analyst, you can create a story to share your ArcGIS Business Analyst sites, maps, and infographics alongside narrative text and other multimedia content. Create your story in ArcGIS Business Analyst and edit and publish it in ArcGIS StoryMaps.

> *Important: This chapter requires a Business Analyst Web App Advanced license.*

# Tutorial 7-1: Create a dashboard

In this tutorial, you will create a site that represents the city center of Portland, Oregon. You will then create a dashboard using the Economic Development Profile and Housing Market Characteristics infographics. You will be able to use the dashboard to view both infographics for your site to gain insight into the economic development and housing opportunities in Portland.

## Create a project

To get started, you need to make a new ArcGIS Business Analyst project.

1. In your browser, navigate to bao.arcgis.com.
2. Enter your ArcGIS credentials and click **Sign In**.
3. On the **Home** tab, click **New project**. Name your project Ch 7 - Portland infographics and click **Create**. Open your project.

   **Tip:** If you are prompted to clear the map of previous data, click **Yes**.

## Set up your map

To create a dashboard, you'll need to have at least one site in your ArcGIS Business Analyst project.

4. Zoom to the general area of Portland, Oregon.

5. Use the **Pin** tool to add a site in the center of Portland.

6. In the pin's pop-up, click **Create site**.

   Point-based sites in Business Analyst Web App are measured with rings, drive times, or walk times. This creates a buffer area around the point.

7. Use the default rings: **1**, **3**, and **5 miles**. Click **Apply**.

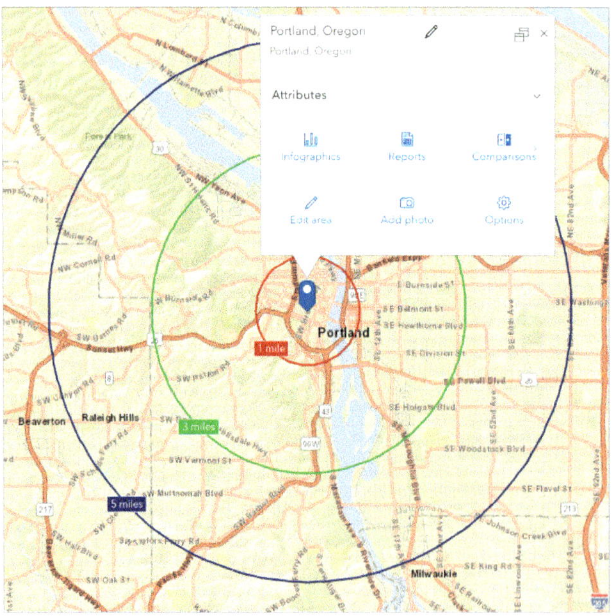


Your site is added to the map.

## Create a dashboard

Now that you have created your site, you can share housing-related information about this site using a dashboard.

8. On the ribbon, click **Share results** and click **ArcGIS Dashboards**.
9. Select the **One site → Different infographics** dashboard type.
10. Select the **Side** layout for the dashboard.
11. Click **Add site** and select the site you recently created representing the city center of Portland, Oregon.
12. Expand the **Standard infographics** section and check the boxes for the **Economic Development Profile** and **Housing Market Characteristics** infographics.
13. Click **Next**.
14. Enter a title for the dashboard, such as Housing in Portland.
15. Click **Create**.
16. When the dashboard is created, click **Open** to view it.

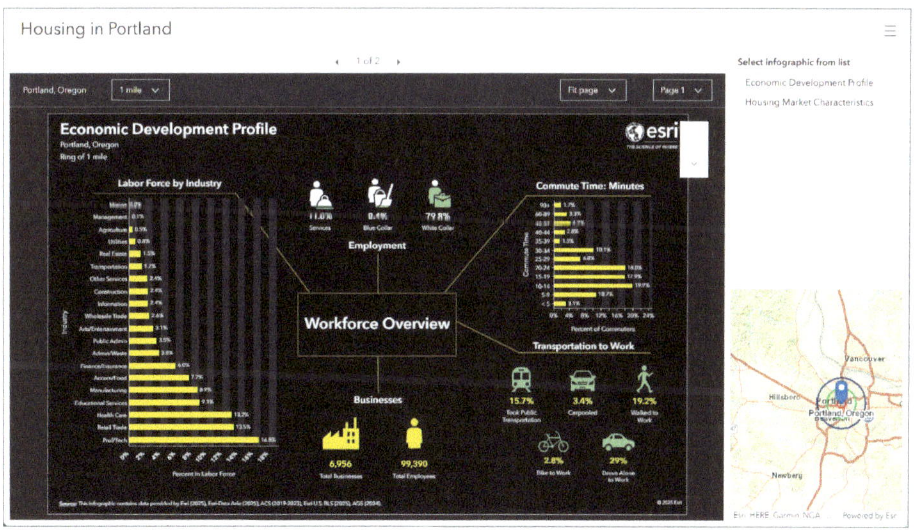


That's it! You've created a dashboard using Business Analyst Web App.

# Tutorial 7-2: Create a story

In the last tutorial, you created a site that represents the city center of Portland, Oregon. You will use that site to create a story using the Housing Market Characteristics infographic. You will be able to use the story to view the infographic and a map of your site to gain insight into the economic development and housing opportunities in Portland.

## Create a project

To get started, you will return to the ArcGIS Business Analyst project from tutorial 7-1.

1. If you are not already in ArcGIS Business Analyst, navigate to bao.arcgis.com and sign in.

2. Open your **Ch** 7 - **Portland infographics** project.

## Set up your map

To create a story, you'll need to have at least one site in your ArcGIS Business Analyst project.

3. On the ribbon, click **Ch** 7 - **Portland infographics** to open the list of sites.

4. If your map is not already zoomed to the Portland site, expand **My point locations** and zoom to the **Portland, Oregon** site.

## Create a story

Now that you have your site, you can share housing-related information about this site using a story.

5. On the ribbon, click **Share results** and click **ArcGIS StoryMaps**.

6. Click **Add sites** and select the **Portland, Oregon** site.

   Now you can configure the items for your story. By default, the boxes for Map, Site photos, Attributes, and Notes are checked.

7. In the **Configure items** window, uncheck the boxes for **Site photos**, **Attributes**, and **Notes.**

8. Check the **Infographic** box.

   Now you'll be able to select an infographic to include in the story.

9. Click **Select infographic** and select the **Housing Market Characteristics** infographic.

10. Click **Apply**.

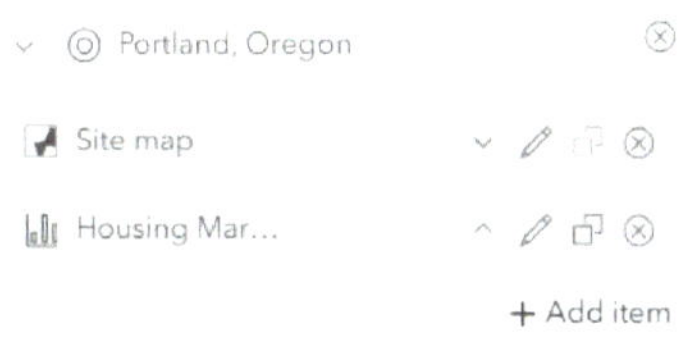


   The infographic element is added to the story.

11. Click **Next**.

12. Keep the default **Floating** selection for the story layout.

13. Enter a title for the story, such as Housing in Portland.

14. Click **Create**.

15. When the story is created, click **Open** to view it.

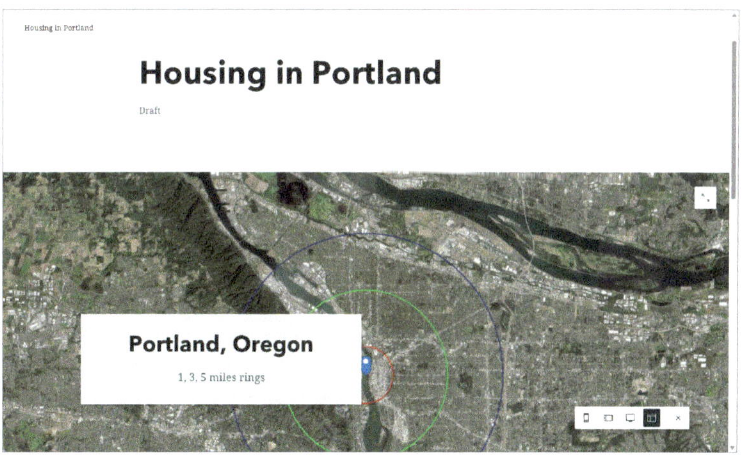

## Summary

In this chapter, you created a site representing the city center of Portland, Oregon, and created a dashboard and a story. With the dashboard, you can view two infographics to better understand the economic development and housing characteristics of the city center. With the story, you can view more details about the housing characteristics and a map of the city center.

### Workflow

1. Create your project.
2. Create a site.
3. Select a dashboard type, layout, and infographic to make a dashboard.
4. Configure your items for your story.

# CHAPTER 8

# Using hexagons in ArcGIS Business Analyst

## Objectives

- Use hexagons to identify at-risk communities with suitability analysis in Business Analyst Web App.
- Use hexagons for cross-border analysis in Business Analyst Pro.
- Enable 3D hexagons and create a 3D scene using hexagons in Business Analyst Pro.

## Introduction

Hexagons, also known as hexbins, are similar to traditional square grids used in GIS, but instead of squares, the grid is made up of six-sided hexagons. In ArcGIS Business Analyst, Uber H3 hexagons are a common method of mapping, as this widely adopted hexagonal grid system covers the entire globe. Each hexagon in the grid is a similar size and has an index ID that never changes—these aspects offer several advantages for analyzing data. H3 hexagons are independent of traditional administrative boundaries and align seamlessly across them, wherever you are in the world. In this chapter, you'll learn the use of hexagons in Business Analyst Web App and Business Analyst Pro.

> *Important: You must have an Advanced license to work with hexagons for mapping and analysis in Business Analyst Web App.*

# Tutorial 8-1: Use hexagons in Business Analyst Web App

In Business Analyst Web App, you can use hexagons in mapping and analysis workflows. In this tutorial, you will perform a suitability analysis to update a response plan for an emergency services organization in Texas City, Texas. You will identify which communities are more at-risk so that the emergency services organization can better prepare for those needs in serving or evacuating that community.

## Create a project

1. In your browser, navigate to bao.arcgis.com.
2. Enter your ArcGIS credentials and click **Sign In**.
3. On the **Home** tab, click **New project**. Name your project Ch 8 - Suitability analysis and click **Create**. Open your project.

 **Tip:** If you are prompted to clear the map of previous data, click **Yes**.

## Perform a suitability analysis

Now you can perform a suitability analysis. The suitability analysis workflow ranks sites according to the criteria you define. You will select your area of interest and hexagon resolution, select your criteria variable list, and modify the calculation methods.

4. Make sure that the **Data source** for the project is set to **USA (Esri 2025)**.

 **Tip:** Hexagons are available only in the US data sources and the Europe (beta) data source.

5. On the ribbon, click **Run analysis** and click **Suitability analysis**.

 For the analysis, you can use your existing sites, features on the map, or standard geographies or hexagons. For this tutorial, you'll be using hexagons.

6. In the **Suitability analysis** pane, for **Select location type**, select **Geographies or hexagons** and click **Next**.

7. For **Define geography**, click the **Area of interest** field and type Texas City-League City CCD, TX and then click the **County Subdivision** result.

8. For the **Level of detail**, click the **Hexagons** tab.

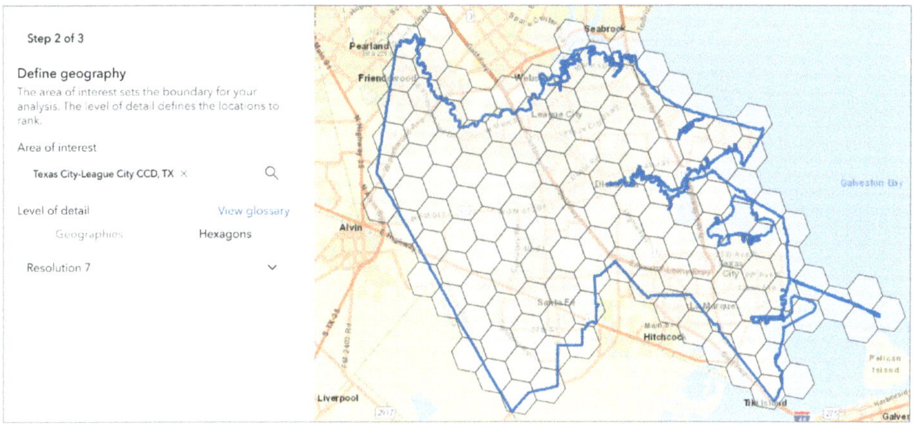


By default, resolution 7 hexagons are used.

> **Tip:** Hexagons can vary in size, from 0 (largest) to 15 (smallest). The size of the hexagon is referred to as the resolution. Using smaller-resolution hexagons (such as resolution 1) is useful for analysis of larger geographic areas, such as country-wide analysis. Bigger-resolution hexagons (such as resolution 15) cover a smaller average area per hexagon, resulting in more hexagons covering the same area.

9. Click **Next** to proceed.

Now you'll select analysis criteria. You can use featured lists of curated variables by industry, your own variable lists, variable lists shared with you by your organization, or browse the data browser to create a new list.

10. Under **Featured lists**, expand the **State and local government** variable list and then click **At-risk**.

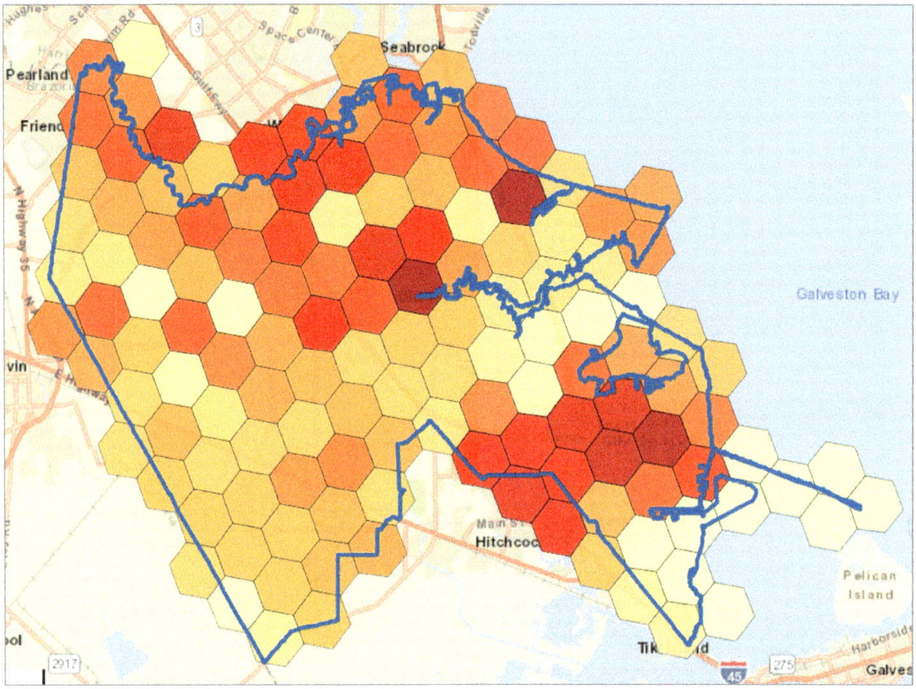


The variables automatically populate on the map layer and in the Results pane.

## Modify the suitability analysis calculations

By default, certain calculation methods are selected. You can modify the influence and range of analysis criteria, select a different scoring method, and adjust criteria weights. In this tutorial, you will change the preset method and final score scale.

11. On the **Suitability analysis** pane, under **Scoring method**, change the **Preset methods** to **Compound differences**.

This preset method is less sensitive to outliers. Therefore, extremely high or extremely low values do not disproportionately impact the overall site score.

12. For **Final score scale**, click **0 to 100**.

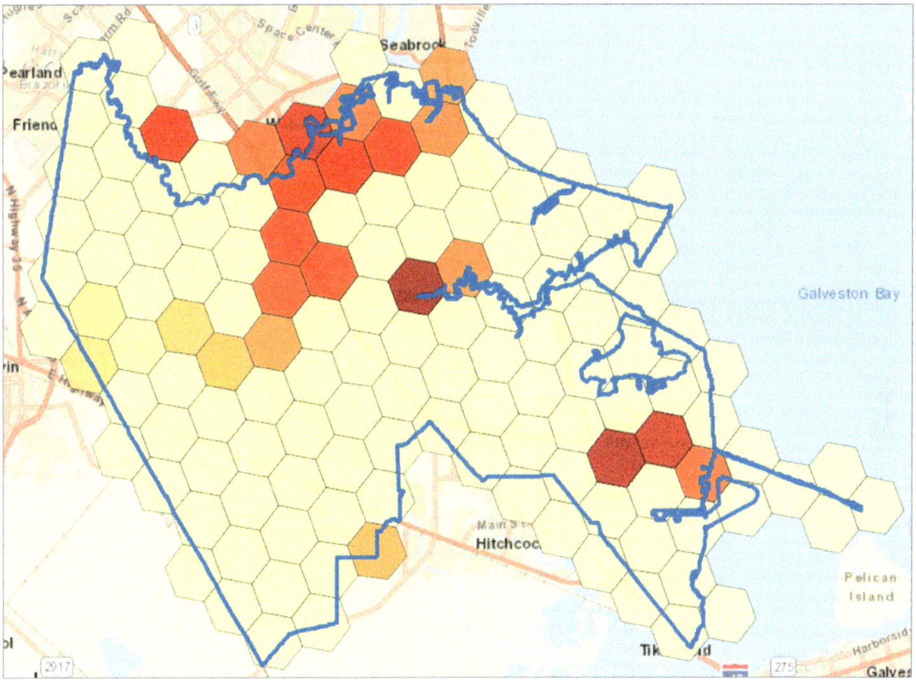


This provides a broader score range that the emergency services department can use to understand the levels of risk for the community.

The map and Results pane are automatically updated. The suitability analysis workflow ranks your hexagons based on the at-risk variable values. A high suitability score indicates a high at-risk score, which is color-coded red on the map.

## Take it to the next level

In the **Results** pane, next to **Hexagon label,** click **ZIP Code** to open the list of label options. Use the geography option to label the hexagons with the name associated with their standard geography at that level of detail. This option automatically adjusts with the hexagon resolution. Because you are using resolution 7 hexagons in your analysis, use the **ZIP Code** option to label the hexagons with the name of their associated zip code.

# Tutorial 8-2: Use hexagons in Business Analyst Pro

In this tutorial, you will use hexagons to represent the total population of countries in Europe. You will then use the 3D capabilities of ArcGIS Pro to add dimension to the hexagon layer.

## Create a project

1. Start ArcGIS Pro. If prompted, sign in using your licensed ArcGIS organizational account.
2. Under **New Project**, click **Map**.
3. In the **New Project** dialog box, apply the following settings:
   - **Name**: Ch 8 - Color-coded hexagons
   - **Location**: C:\Top20BusinessAnalyst

   **Tip:** This book uses **C:\Top20BusinessAnalyst** for the folder location, but you can use any folder.

4. Click **OK** to create your project.

## Set the data source

5. On the ribbon, click the **Analysis** tab. In the **Workflows** group, click **Business Analysis**, and at the bottom of the menu, click **Change data source**.
6. In the **Business Analyst Data Source** window, under **Portal**, click **Europe** and then click the **Europe** data source.
7. Click **OK**.

   This dataset contains data for 39 European countries and allows you to perform cross-border mapping and analysis.

## Map the population in Europe

8. Open the **Business Analysis** menu. Under **General Workflows**, click **Color Coded Layer**.

   The Data Browser window appears, allowing you to select a variable to map.

9. In the **Data Browser** window, click the **Population** card and then click the **Common Population Variables** card. Select **2024 Total Population**.

10. Click **OK**.

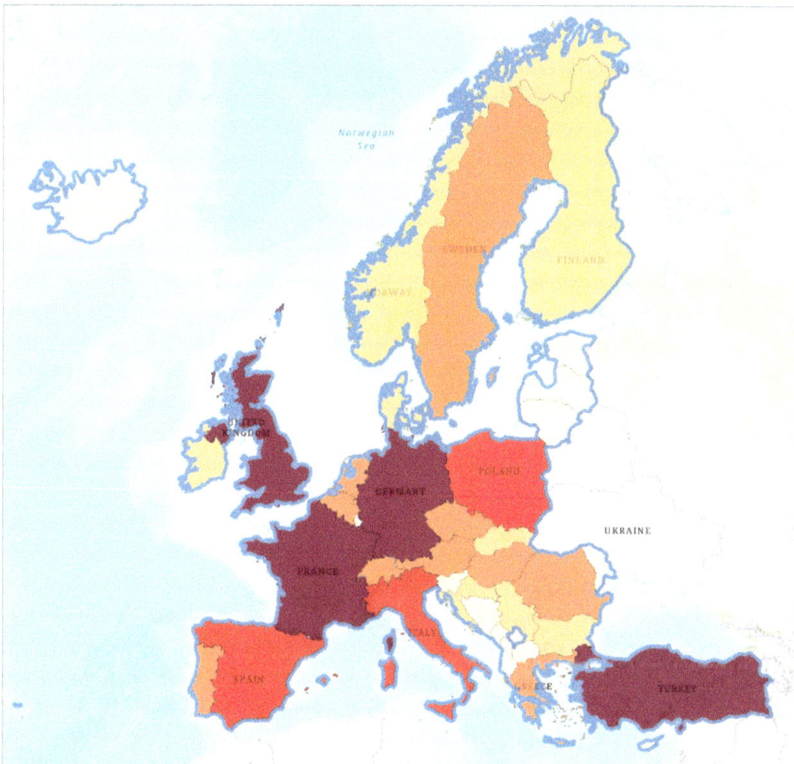


> **Tip:** If your map does not automatically zoom to the layer, in the **Contents** pane, right-click **Color Coded Layer** and click **Zoom To Layer**.

The variable is mapped for the 39 European countries. Next, you'll set a specific area of interest within Europe.

11. In the **Symbology** pane, under **Geography**, in the **Area of interest** field, type Estonia. Click the suggested location, **Estonia (EE)**.

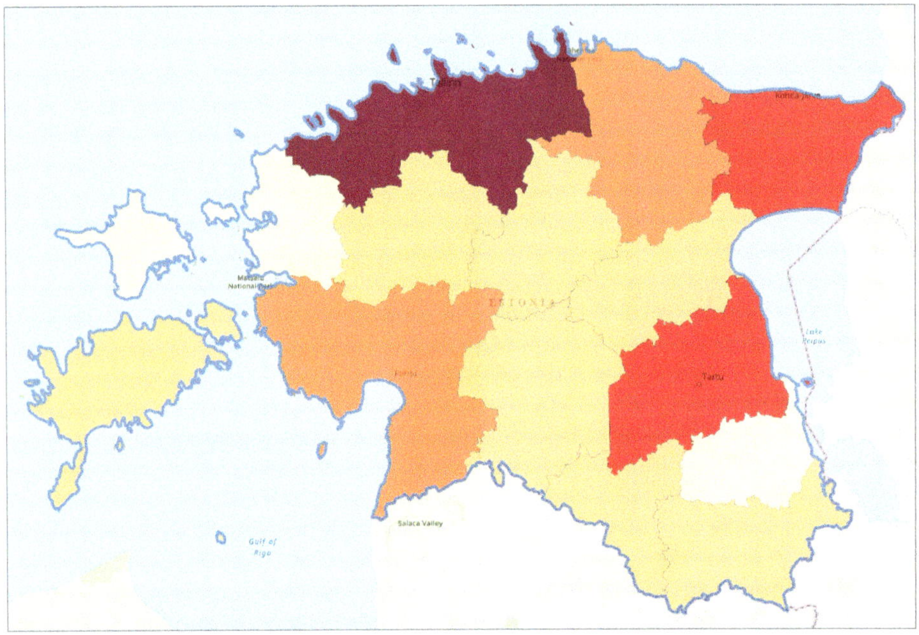

The map immediately updates to show 2024 Total Population for Estonia. Now you'll add two more countries to the area of interest.

12. Repeat the previous step, adding the countries of Latvia and Lithuania to the area of interest.

The map now shows the 2024 Total Population variable for the three Baltic states. The level of detail is automatically set to Major Subdivisions, a common geographic boundary in Europe. Next, you'll change this level to resolution 6 hexagons.

## Style the map with hexagons

13. In the **Symbology** pane, for **Boundary mode**, click **Hexagons**.

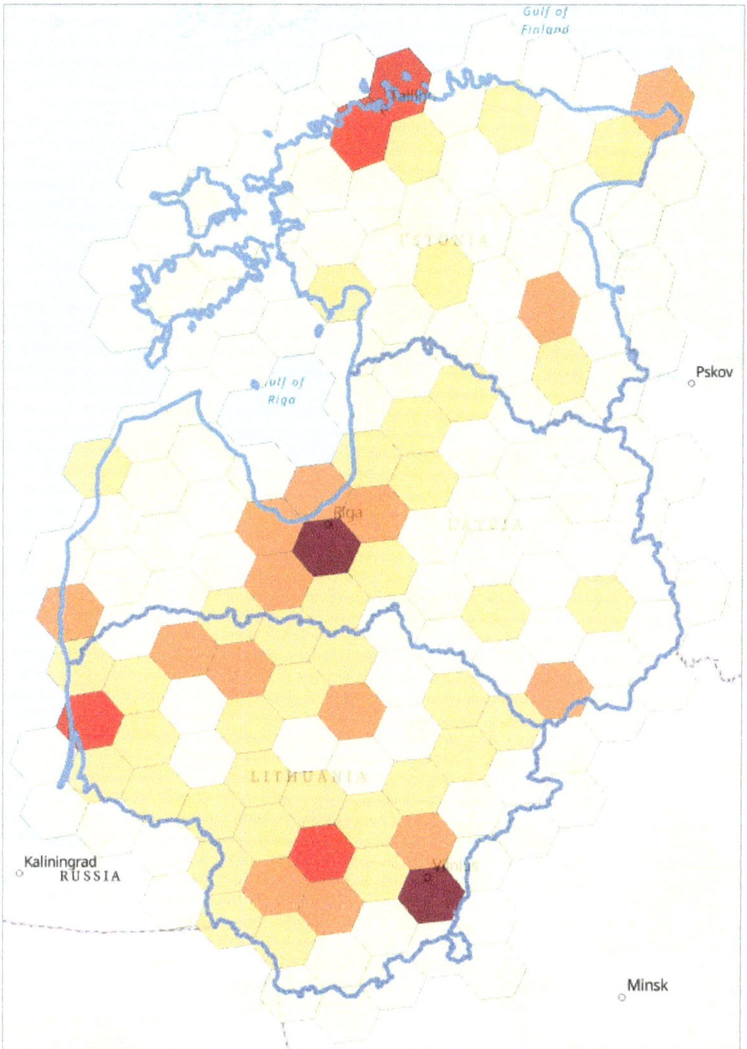


The map refreshes to show the population of the three countries, mapped in resolution 4 hexagons.

14. For **Level of detail**, click **Resolution 6**.

    The map displays the same data, mapped in smaller-size hexagons. This allows population centers to stand out but doesn't show much information about the areas outside major cities. Next, you'll break the data into a larger number of classes.

15. Under **Style**, for **Classes**, click **30**.

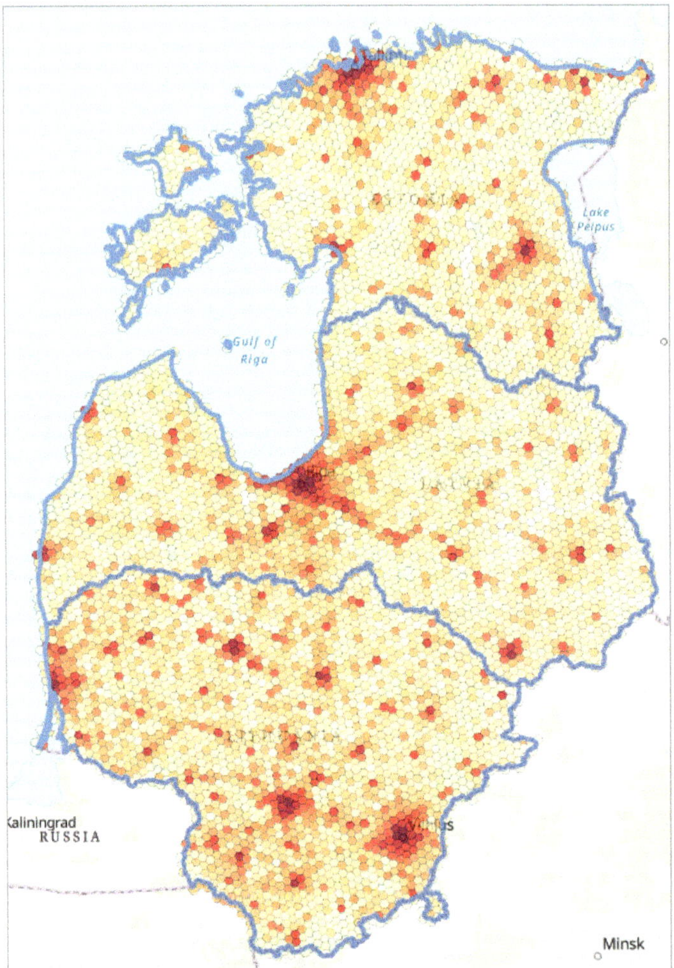


The map updates, and you can now see a more textured breakdown of population across the three countries.

## Enable 3D hexagons

16. Click the **Analysis** tab. In the **Geoprocessing** group, click **Tools**. Search for Feature to 3D By Attribute and open the tool.

17. In the **Feature to 3D By Attribute** tool, apply the following settings:

    - For **Input Features**, select **Color Coded Layer/Resolution 6**.
    - For **Output Feature Class**, type TotPop2024_3D.
    - For **Height Field**, select **2024 Total Population**.
    - Leave **To Height Field** blank.

18. Click **Run**.

    A new layer appears on the map, containing the 2024 Total Population variable as a height field for 3D data display. However, this map is not 3D so you will now create a 3D scene.

## Create a 3D scene using hexagons

19. For ease of viewing, in the **Contents** pane, check the box next to **Color Coded Layer** to turn off its visibility.

20. On the ribbon, click the **Insert** tab. In the **Project** group, click the **New Map** drop-down menu. Click **New Local Scene**.

    Your project now has two map views: a 2D map and a 3D scene.

21. Click the **Map** view tab. In the **Contents** pane, right-click the **TotPop2024_3D** layer and click **Copy**.

22. Click the **Scene** view tab. In the **Contents** pane, right-click **Scene** and click **Paste**.

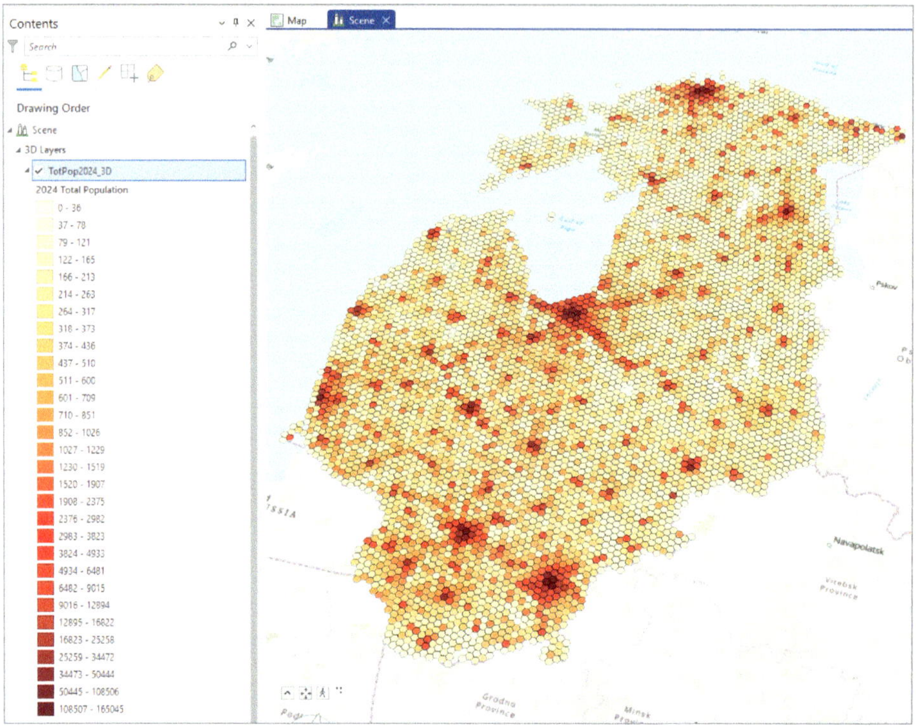


The 3D-enabled hexagon layer is now in the scene. You'll now adjust settings to visualize the hexagons in 3D.

23. In the **Contents** pane, right-click the 3D layer and click **Properties**.

24. In the **Layer Properties** window, click the **Elevation** tab. Open the **Features are** drop-down menu and click **On the ground**. Click **OK**.

25. In the **Contents** pane, make sure the 3D layer is selected. On the ribbon, click the **Feature Layer** tab.

26. In the **Extrusion** group, apply the following settings:

- For **Type**, select **Base Height**.
- For **Field**, select **2024 Total Population**.

27. Use the joystick at the lower left of the map to navigate the 3D scene, which displays resolution 6 hexagons mapping total population in Estonia, Latvia, and Lithuania.

# Summary

In this chapter, you used hexagons as a geography option for a suitability analysis in Business Analyst Web App. You also created a layer of color-coded hexagons in Business Analyst Pro and used them to create a 3D scene.

## Workflow

1. Open a workflow.
2. Set the geography type to hexagons.
3. Specify the hexagon resolution.
4. Use hexagons for analysis.

# CHAPTER 9

# Using smart map search

Coauthor Elif Bulut

## Objectives

- Add a layer to the map for use as smart map search criteria.
- Select county health data as variables for a smart map search.
- Explore the results and summary information.

## Introduction

Smart map search allows you to view geography areas that match your criteria. You can select up to 10 variables, set ranges, and view the results on the map and in the Results pane. You can export the results to Excel or click any geography on the map for more options.

# Tutorial 9-1: Use smart map search in Business Analyst Web App

In this tutorial, you will explore which counties in the United States have a lower life expectancy than the national average, which was 76.4 years at the time of writing.

## Create a project and set up smart map search

To start this tutorial, you'll need to create a new project and add a web map of County Health Rankings (CHR) data. CHR data collects average life expectancy statistics for each group at the county level, providing a quick overview of general trends in life expectancy across demographic groups in the United States.

1. In your browser, navigate to bao.arcgis.com.
2. Enter your ArcGIS credentials and click **Sign In**.
3. On the **Home** tab, click **New project**. Name your project Ch 9 - US life expectancy and click **Create**. Open your project.

   **Tip:** If you are prompted to clear the map of previous data, click **Yes**.

4. On the **Map** tab, click **Add data** and click **Web maps and layers**.
5. In the **Web maps and layers** window, under **Explore**, expand the **Living Atlas** group and click **Show all**.
6. In the search bar, type County Health Rankings 2025. In the search results, hover over the **Country Health Rankings 2025** map and click **Add to map**.

   The map now displays the CHR map, and its data is now saved to the project. The map is color-coded by county, using a dark-purple to light-yellow color ramp. Each county's color corresponds to the percentage of adults in that county that report fair or poor health. It's an immediately legible and compelling visualization—but did you know that the CHR data contains more than the web map? It also has a lot of health-related variables.

7. To start using smart map search, on the ribbon, click **Create maps** and click **Smart map search**.

8. In the **Smart map search** pane, expand the **Geography** section. For **Area of interest**, type United States and click **USA (USA)**. For **Level of detail**, open the drop-down menu and click **Counties**.

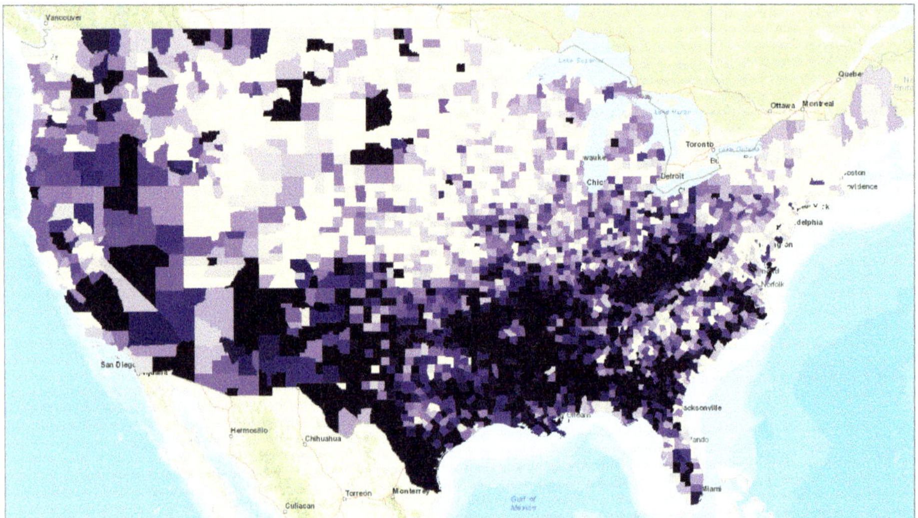


**Tip:** Click **View glossary** to learn more about the definitions of each geographic level.

## Add variables to the smart map search

Now you can select which variables to use as search criteria.

9. At the bottom of the pane, click **Browse all variables** to use the **Data browser** window.

10. Click the **Other variables** tab. Under the search bar, select the **Map layers** category.

11. Expand the **County (Map Layers)** section. Scroll down and add the following variables:

    - **Average expected number of years of life from birth** (for each of the six race groups)
    - **Median household income**
    - **Percentage of the population food insecure**

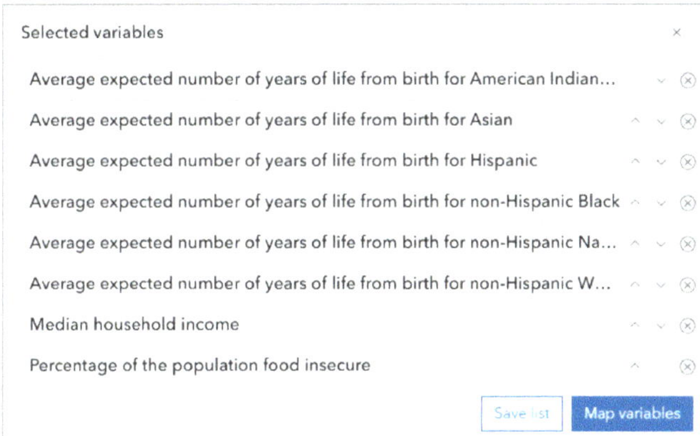


You should have eight selected variables.

12. Click **Apply.**

**Tip:** If you encounter a geography warning, select the option **Use the selected variables**. Click **Confirm**.

Now that you've selected variables, they appear in the workflow pane with their ranges displayed. Because these are long variable names that get truncated, you can always hover over a variable to see the full name.

13. In the **Smart map search** pane, under **Variable list**, for each **Average expected number of years of life from birth** variable, change the minimum value to 1 and the maximum value for each to 99.

14. Scroll down to the **Style** section. Click **Show style options**. Under **Matching areas**, change the **Transparency** value to 0.

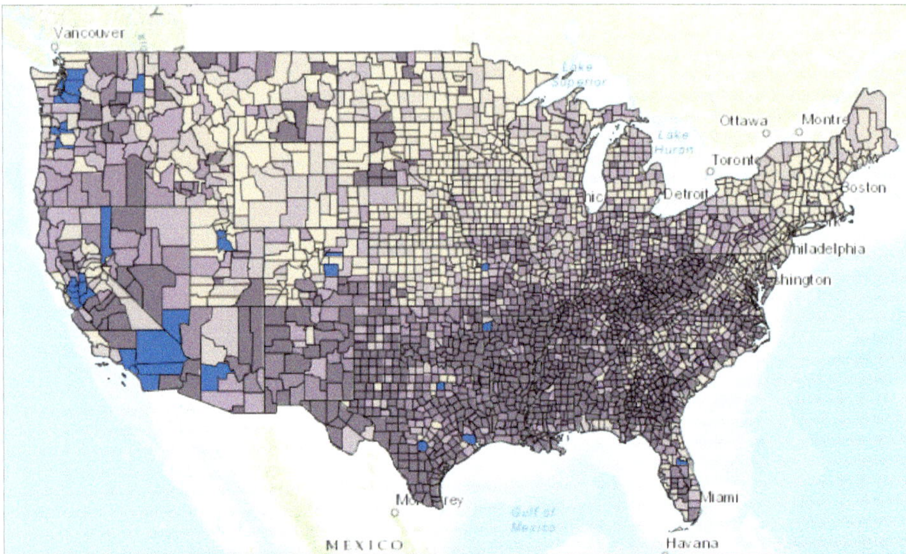


Filtering to exclude missing values and outliers helps normalize the data and improve the accuracy of the analysis. Using the range 1–99, any values of zero are removed. Typically, when a county does not have reported data for a health variable, a value of 0 is assigned. When the zero values are removed, the map highlights only counties that reported data about life expectancy. These counties are shown in blue.

## Explore the results

Now you can explore the results on the map and in the interactive Results pane. In the Results pane, explore the smart map search results as a summary, histogram, bubble chart or scatterplot, and table.

15. Click the **Summary** tab to view a summary of the results, such as the number and type of variables you mapped. The summary includes the count of areas, average value of the variable calculation, the top five and bottom five areas, a statistical trends chart, and a relationship chart.

16. Click the **Histogram** tab to view the results for each variable in a histogram.
17. Click the **Bubble chart** tab to view the results for each variable as a bubble chart or scatterplot.
18. Click the **Table** tab to view the results in a table and optionally export the standard geography data to an Excel spreadsheet.

## Take it to the next level

You have mapped life expectancy for counties in the United States. You can explore which counties in the United States have a lower life expectancy than the current national average of 76.4 years in the interactive **Results** pane.

On the **Summary** tab, you can see the average life expectancy for each race group in the counties that reported data. Click the **Within range** hover option to display areas on the map that have data for the variable and areas that do not.

On the **Histogram** tab, explore places where specific groups exhibit notably high or low life expectancies. Use the variable **Average expected number of years of life from birth for non-Hispanic Blacks** to show the data for just this race group, helping you explore areas with the highest and lowest life expectancy for the group.

On the **Bubble chart** tab, choose **Average expected years of life from birth for non-Hispanic Blacks** for the y-axis, **Median household income** for the x-axis, and use **Percentage of population who are food insecure** to determine the dot size. In this example, you can notice larger bubbles on the lower left of the chart, indicating that areas with more food-insecure residents tend to have lower household income and reduced life expectancy.

On the **Bubble chart** tab, use the **Scatterplot** option. Choose **Average expected number of years of life from birth for non-Hispanic whites** for the y-axis. You can observe that median household income accounts for 55 percent of the variation in life expectancy. On the other hand, when you select **Average expected number of years of life from birth for non-Hispanic Blacks** for the y-axis, the regression line for Black individuals appears to be less steep and you can observe that median household income explains a smaller percentage of variation in life expectancy for this group than non-Hispanic whites.

## Summary

In this chapter, you used smart map search in Business Analyst Web App.

### Workflow

1. Open your project and add a web map.
2. Add variables to the smart map search.
3. Use filters to specify your search.
4. Explore the results.

# CHAPTER 10

# Running a suitability analysis

## Coauthor Elif Bulut

## Objectives

- Choose and map sites for analysis.
- Select and refine suitability criteria.
- Use the Results pane to explore the analysis.

## Introduction

As a discipline, GIS focuses on digitally mapping and analyzing locations. But what if you need to choose a location? How do you make decisions based on the geographic system you have created in ArcGIS? This is where ArcGIS Business Analyst gets to flex its muscles.

As you learned in chapter 9, ArcGIS Business Analyst uses demographic data to help users understand the qualitative and quantitative differences between locations. A smart map search shows you all the locations that meet the variable criteria that matter to you. This chapter focuses on suitability analysis, a workflow that distinguishes locations by ranking them according to your criteria.

How you define a "suitable" location will vary across use cases. In some cases, such as opening a new business location, suitable areas may be places teeming with your ideal customers. At other times, you might define suitability as a high need for a service or policy intervention. In this chapter, we'll use an example of both types: finding the most customers for a new urgent care center with Business Analyst Web App and locating the most socioeconomically vulnerable populations using Business Analyst Pro.

## Tutorial 10-1: Perform a suitability analysis in Business Analyst Web App

In this tutorial, you will use suitability analysis in Business Analyst Web App to locate a site for a new urgent care center. You'll create a point layer of existing center locations to use as a suitability criterion, in addition to demographic variables that approximate a profile of potential urgent care center patients.

### Create a project

1. In your browser, navigate to bao.arcgis.com.
2. Enter your ArcGIS credentials and click **Sign In**.
3. On the **Home** tab, click **New project**. Name your project Ch 10 - Age dependency suitability analysis and click **Create**. Open your project.

   **Tip:** If you are prompted to clear the map of previous data, click **Yes**.

### Select sites

4. On the ribbon, click **Run analysis** and click **Suitability analysis**.
5. In the **Suitability analysis** pane, for the location type, click **Geographies**. Click **Next**.

   You'll use block groups to rank potential locations for a new urgent care facility.
6. Click the **Area of interest** field and type Maverick County, TX. Select **Maverick County, TX (County)** to lock the analysis extent to this area.

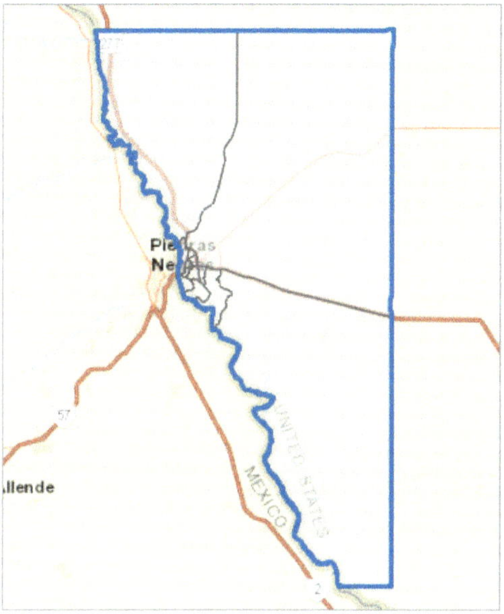


7. Change the **Level of detail** to **Block groups** as the geographies you'll use in your analysis, and then click **Next**.

## Select criteria

8. At the bottom of the **Suitability analysis** pane, click **Select criteria** and select **Add variables from data browser**.

9. In the **Data browser** window, under **Standard variables**, click the **Population** card.

   All population-related data is displayed, with the most popular variables at the top, allowing you to browse the options.

10. Check the box next to **2025 Population Density (Pop per Square Mile) (Esri)**.

    **Tip:** If the 2025 dataset is not available, use the most recent dataset.

    The variable is added to the Selected variables list. You can access and edit the list at any time during the data browser session.

11. Click **Explore** to return to the main **Data browser** page.

12. In the search bar, type age dependency ratio. Once the search results appear, check the box next to **2025 Age Dependency Ratio (Esri)** to add it to the variable list.

13. Use a combination of browsing and searching to locate the remaining 11 variables on this list:

    - **2025-2030 Population: Compound Annual Growth Rate (Esri)**
    - **2025 Ambulatory Health Care Services (NAICS621) Businesses**
    - **2025 Median Household Income (Esri)**
    - **2023 Population w/Income Below Poverty Level (ACS 5-Yr)**
    - **2023 Pop <19: No Health Insurance Coverage (ACS 5-Yr)**
    - **2023 Pop 19-34: No Health Insurance Coverage (ACS 5-Yr)**
    - **2023 Pop 35-64: No Health Insurance Coverage (ACS 5-Yr)**
    - **2023 Pop 65+: No Health Insurance Coverage (ACS 5-Yr)**
    - **2025 Important to See Doctor When Sick: 4-Agree Completely**
    - **2025 Prefer Urgent Care vs. Doctor's Office: 4-Agree Completely**
    - **2025 Struggle with Health Care Costs: 4-Agree Completely**

14. On the top right of the window, click the **Selected variables** list to double-check the 13 variables you added. Click **Save list** and name the list Urgent care variables. Click **Save**.

15. At the bottom of the window, click **Apply**.

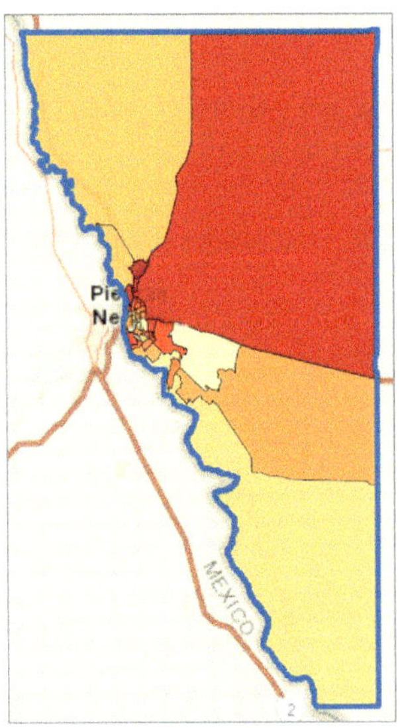

The suitability analysis workflow automatically runs. Each block group in Maverick County is shaded to show the degree of suitability according to your criteria, and the Results pane displays a summary of the results.

## Adjust criteria

16. In the **Suitability analysis** pane, scroll down the list of **Analysis criteria** to locate **2025 Ambulatory Health Care Bus (NAICS)**. Under **Influence**, select **Inverse**.

    As the number of urgent care centers in an area rises, the demand for an additional center decreases, inversely affecting the need for a new one. Setting the influence to inverse means that the fewer health-care businesses in the block group, the higher the block group is ranked in the analysis.

17. Scroll down the **Suitability analysis** pane to the **Weighting** section. Click **Adjust weights**, set the locate **2025 Ambulatory Health Care Bus (NAICS)** criterion weight to 15%, and click the lock button.

    A higher weight indicates that this criterion is more important for your analysis than the others.

18. In the **Adjust weights** window, click **Done**.

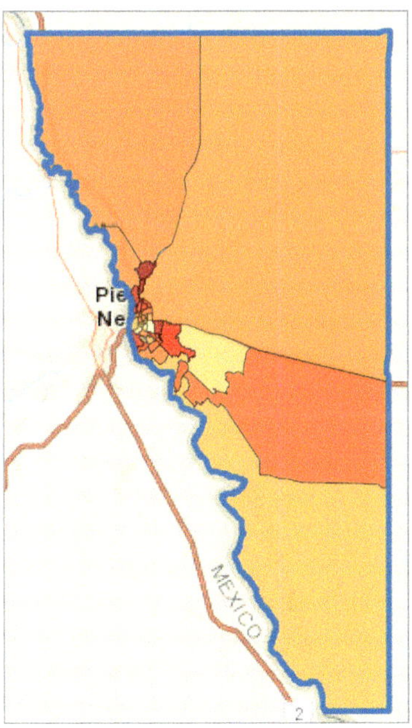

## Explore the results

19. In the **Results** pane, hover over each bar in the **Top 5** summary card to view the most suitable block groups for a new urgent care center.

    As you hover over a bar, the corresponding block group will be highlighted on the map.

20. In the **Suitability analysis** pane, scroll down to the **Legend** section and click **Filter results**. Select **Filter by rank** and leave the default of the top five locations, and then click **Done**.

21. On the map, zoom in to the filtered locations.

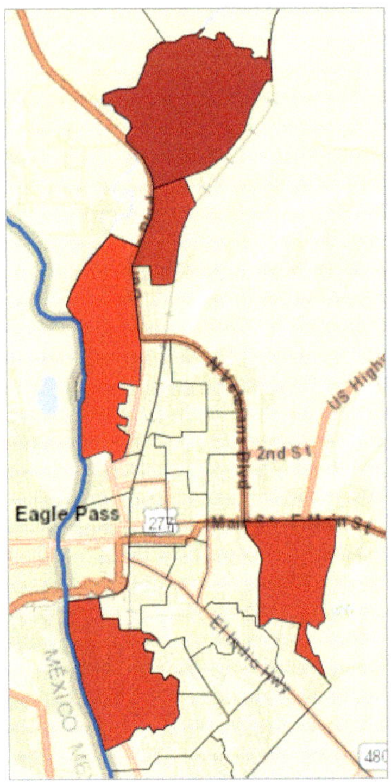


    The map now shows only the block groups in Maverick County that are most suitable for a new urgent care center.

22. In the **Results** pane, explore the analysis results using the **Histogram**, **Bubble chart**, and **Table** tabs.

### Take it to the next level

Your analysis has identified the most suitable block group for a new urgent care location, based on your criteria. To learn more about this block group, you could create an infographic in ArcGIS Business Analyst. In chapter 3, you learned that an infographic presents a set of important information about a chosen location.

First, locate the top-ranked block group in your analysis and click on it in the map. In the pop-up, click **Infographics**. This launches the infographic player, which allows you to run any of the ArcGIS Business Analyst infographics for the location. In the upper left, in the **Infographic** drop-down list, expand **Gallery infographics** and select **Health Care for At-Risk Populations**.

This infographic template provides a comprehensive overview of health-care factors that affect at-risk populations. Variables included in the template present information on hospital locations, income, health-care habits, health insurance, and medical spending. This infographic may be useful for those considering opening a new health-care business in a specific location.

## Tutorial 10-2: Perform a suitability analysis in Business Analyst Pro

In this tutorial, you will be performing a suitability analysis on the socioeconomic vulnerability in Gwinnett County, Georgia, using Business Analyst Pro.

### Create a project

1. Start ArcGIS Pro. If prompted, sign in using your licensed ArcGIS organizational account.
2. Under **New Project**, click **Map**.
3. In the **New Project** dialog box, apply the following settings:
   - **Name**: Ch 10 - Socioeconomic vulnerability
   - **Location**: C:\Top20BusinessAnalyst

**Tip:** This book uses **C:\Top20BusinessAnalyst** for the folder location, but you can use any folder.

4. Click **OK** to create your project.

## Add locations to the map

5. On the ribbon, click the **Analysis** tab. In the **Geoprocessing** group, click **Tools**.

6. In the **Geoprocessing** pane, search for Generate Standard Geography Trade Areas and open the tool.

   Standard geography trade areas are administrative and political boundaries, such as states, counties, and zip codes.

7. In the **Generate Standard Geography Trade Areas** tool, apply the following settings:

   - For **Geography Level**, select **Census Tracts (US.Tracts)**.
   - For **Output Feature Class**, type GwinnettCounty_TA.
   - For **Input Type**, select **List**.
   - For **Geography IDs List**, click the browse button. In the **Select Geography: US.Tracts** window, select **US by Tracts** and then select **Georgia**. In the list of counties, scroll down and select **Gwinnett County**. Check the box for **Gwinnett County** to select all 220 census tracts in the county and click **OK**.

8. At the bottom of the tool, click **Run**.

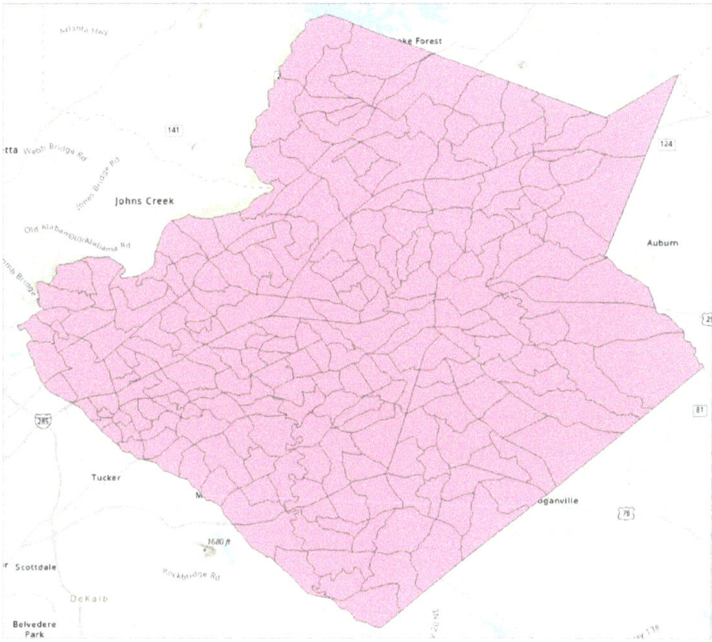


The tool runs and a layer of census tracts in Gwinnett County is added to the map.

## Select candidate sites

9. On the **Analysis** tab, in the **Workflows** group, open the **Business Analysis** drop-down menu. Under **General Workflows**, click **Suitability Analysis**.

10. In the **Make Suitability Analysis Layer** window, for **Input Features**, click **GwinnettCounty_TA** and click **OK**.

    A new suitability analysis layer is created and added to the Contents pane.

## Select criteria

11. On the ribbon, click the **Suitability Analysis** contextual tab that just appeared.

    This opens a workflow-specific ribbon that contains all the tools for suitability analysis. The tab and ribbon are accessible only when you have selected a suitability analysis layer in the Contents pane.

12. In the **Criteria** group, click the arrow below **Add Criteria** and click **Add Variables From Data Browser**.

13. In the **Add Variable Based Suitability Criteria** window, next to **Variables**, click the plus button.

    The Data Browser window appears.

14. Use a combination of browsing and searching to locate the following 12 variables, being careful to check the box for each one and select the correct calculation type:

    - **2025 Median Household Income (Esri) (#)**
    - **2023 Households Receiving Food Stamps/SNAP (ACS 5-Yr) (%)**
    - **2025 Median Home Value (Esri) (#)**
    - **2025 Owner Occupied Housing Units (Esri) (%)**
    - **2023 HHs w/Gross Rent 50% + of Household Income (ACS 5-Yr) (%)**
    - **2023 Household Pop w/Income Below Poverty Level (ACS 5-Yr) (%)**
    - **2025 Population Age 25+: Bachelor's Degree (Esri) (%)**
    - **2025 Unemployed Population Age 16+ (Esri) (%)**
    - **2023 Households with 1+ Persons with a Disability (ACS 5-Yr) (%)**
    - **2023 Population 35-64: No Health Insurance Coverage (ACS 5-Yr) (%)**
    - **2023 Population 65+: No Health Insurance Coverage (ACS 5-Yr) (%)**
    - **2023 HHs with No Internet Access (ACS 5-Yr) (%)**

15. Below the search bar, click the **Show/Hide selected variables** button.

16. In the **Selected Variables** pane, double-check the variables you added. Click **Save List** and name the list Socioeconomic vulnerability variables. Click **OK**.

17. At the bottom of the **Add Variable Based Suitability Criteria** window, click **OK**.

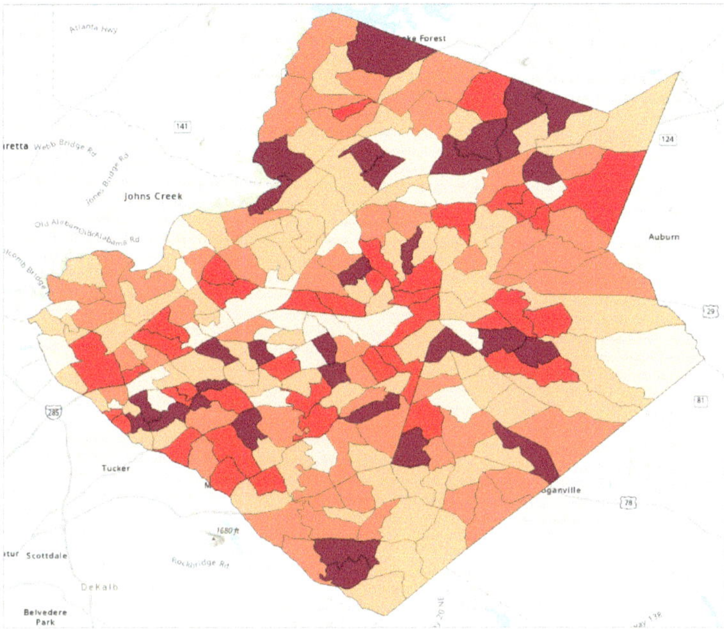


The analysis runs. It calculates a suitability score for each census tract based on the fields you selected. The suitability analysis layer symbology updates to show the range of the suitability score values.

## Adjust criteria

18. In the **Suitability analysis** pane, for each of the following criteria, change the influence setting to **Inverse**:

- **2025 Median Household Income**
- **2025 Median Home Value**
- **2025 Owner Occupied HUs: Percent**
- **2025 Pop Age 25+: Bachelor's Degree: Percent**

As discussed in the previous section, setting the influence to inverse for these criteria indicates that a higher value for these variables indicates a lower degree of socioeconomic vulnerability.

19. At the top of the pane, click the **Settings** tab. Go to the **Weighting** section.

20. Adjust the weights for the following variables using the given values. Click the lock button after changing the value:

- **2025 Median Household Income**: 10.00%
- **2023 HHs w/Food Stamps/SNAP (ACS 5-Yr)**: 3.00%

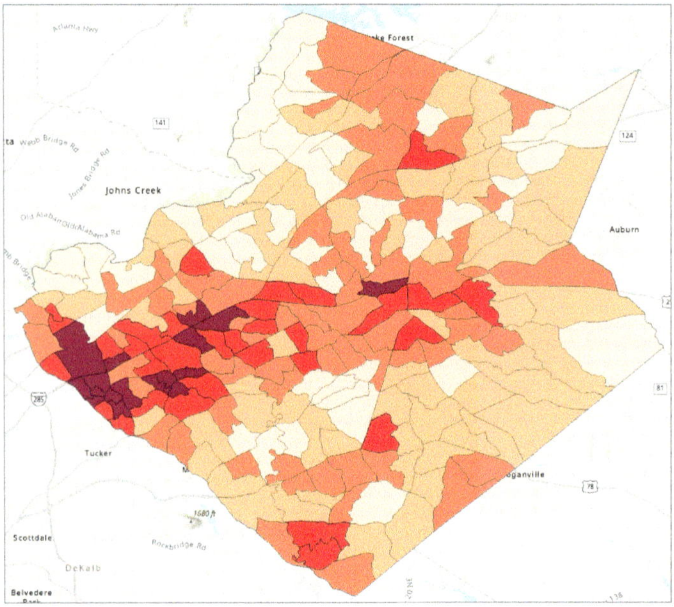


## Explore the results

21. At the bottom of the **Suitability analysis** pane, click **View Results** to open the **Results** pane.

22. In the **Results** pane, in the **Top 5** summary card, click each bar to view the most suitable census tracts for policy intervention aimed at socioeconomic assistance.

23. In the **Suitability analysis** pane, on the **Settings** tab, go to the **Scoring** section.

24. Turn on **Enable filtering**. Select **Filter by Rank** and use the drop-down lists to specify **Top 5**.

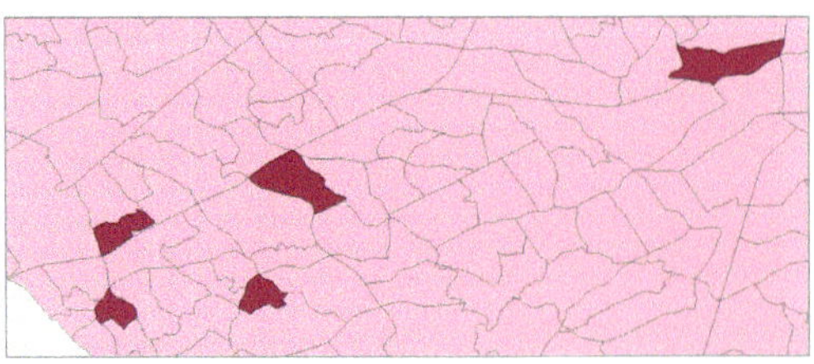

The map now shows only the census tracts in Gwinnett County that are most in need of intervention, based on your socioeconomic criteria.

25. In the **Results** pane, explore the analysis results using the **Histogram**, **Bubble chart**, and **Table** tabs.

### Take it to the next level

Race and ethnicity are often intertwined with socioeconomic vulnerability and can be important factors to consider when determining how to prioritize resources for intervention. To contextualize your analysis results, you can add an ArcGIS Living Atlas layer showing the predominant race in each census tract.

1. On the ribbon, click the **Map** tab. In the **Layer** group, click the **Add Data** button.
2. In the **Add Data** window, under **Portal**, click **Living Atlas**. Search for ACS Race and Hispanic Origin Variables.
3. Double-click **ACS Race and Hispanic Origin Variables - Centroids** and then double-click the **Tract** layer.

The layer shows centroids, or point symbols that represent the center of a polygon (in this case, census tract polygons). The size of the circle corresponds to population, whereas the color corresponds to the predominant race or ethnicity in each tract. The transparency of each symbol indicates the strength of predominance.

## Summary

In this chapter, you used suitability analysis in Business Analyst Web App to locate areas with the most potential customers for a new urgent care center. You also used suitability analysis in Business Analyst Pro to locate socioeconomically vulnerable populations.

### Workflow

1. Select sites for analysis.
2. Select criteria to evaluate the sites.
3. Adjust the criteria using weights and influence.
4. Analyze the results.

# CHAPTER 11

# Comparing and analyzing sites using benchmark comparisons

Coauthors Elif Bulut, Jared Knight, and Ryan Phillips

## Objectives

- Identify sites and variables to compare.
- Use different levels of geographies to perform various types of analysis.
- Gain insights on the surrounding area of sites.

## Introduction

The benchmark comparisons workflow compares sites using criteria and benchmark values that you define. First, select sites to include in the analysis, and then choose variables and set a benchmark for comparison. The workflow compares your sites—displayed as color-coded symbols on the map—and adjusts the analysis as you change the comparison method or benchmark value. The results of this analysis appear in the Results pane and can be saved as a new layer in your project or exported to an Excel worksheet.

In this chapter, you'll run benchmark comparisons in Business Analyst Web App and Business Analyst Pro.

# Tutorial 11-1: Run benchmark comparisons in Business Analyst Web App

In this tutorial, you will compare municipalities in Kyoto, Japan. You are a real estate developer analyzing projected household growth to determine where to build your next housing development.

## Create a project

You will open your project and set your data source.

1. In your browser, navigate to bao.arcgis.com.
2. Enter your ArcGIS credentials and click **Sign In**.
3. On the **Home** tab, click **New project**. Name your project Ch 11 - Benchmark comparisons and click **Create**. Open your project.

   **Tip:** If you are prompted to clear the map of previous data, click **Yes**.

4. At the top of the page, click **Data source** to open the list of data sources.
5. Scroll to find **Japan** or enter Japan in the search field. Click the **Click for data source selections** button. Next to **Esri Japan**, click **Apply**.

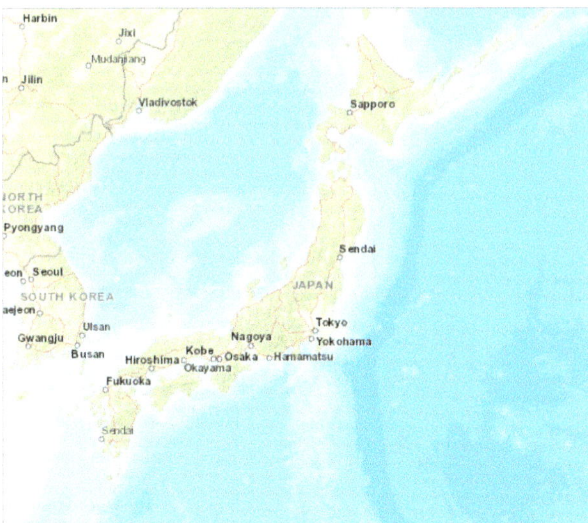


The map adjusts so that Japan is the area of interest.

## Set the area of interest

For this tutorial, you'll use standard geographies, with Kyoto as the area of interest and municipalities as the level of detail.

6. On the ribbon, click **Run analysis** and click **Benchmark comparisons**.

   The Benchmark comparisons pane appears. For the location type, you can select an existing site in your project, features on the map, or standard geographies or hexagons.

7. In the **Benchmark comparisons** pane, for the location type, click **Geographies** and click **Next**.

8. Click the **Area of interest** field and type Kyoto. Select the **26 Kyoto (Prefecture)** option.

9. Click the **Level of detail** to open the list and click **Municipalities**, if not selected already.

10. Click **Next**.

## Select analysis criteria

Benchmark comparisons use criteria lists to compare variables across sites. You can use curated variable lists, custom lists that you create, or shared lists.

11. Under the **Popular** lists, next to the **Households list**, click the **Show list** button.

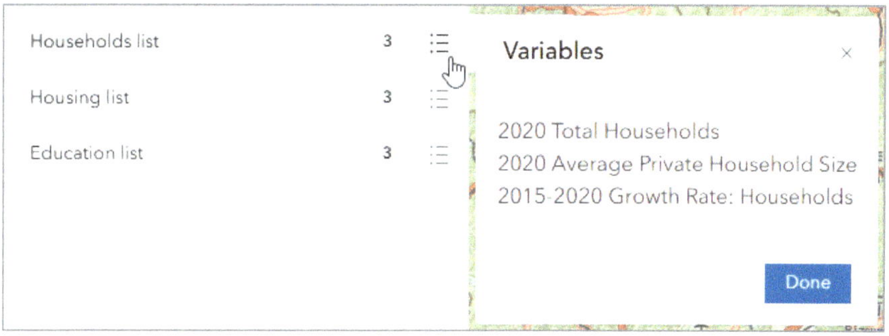


The variables included in this featured list are displayed. This list includes three demographic variables about total households, average private household size, and projected household growth rate.

12. Click **Households list**.

13. On the map, hover over a municipality.

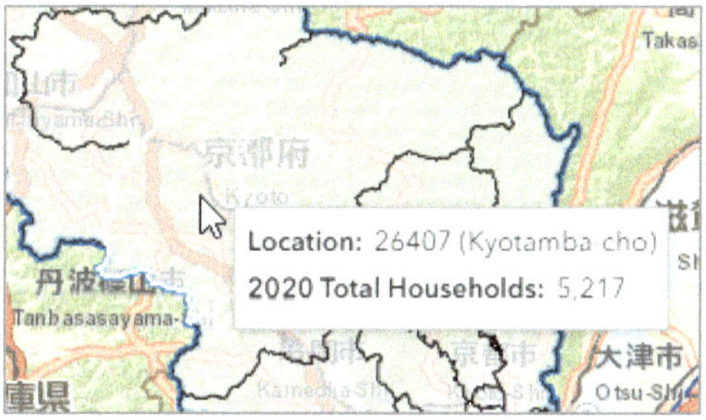


When you hover over each municipality, the results are displayed on the map. They are also in the Results pane.

14. In the **Benchmark comparisons**, in the **Statistics** section, turn on the option for **Mean**.

You can enable statistics (including mean, median, and standard deviation) for the table. Mean or median can be used as the benchmark value.

## Explore the results

15. In the **Results** pane, next to **Mean**, click the three dots. Click **Make benchmark**.

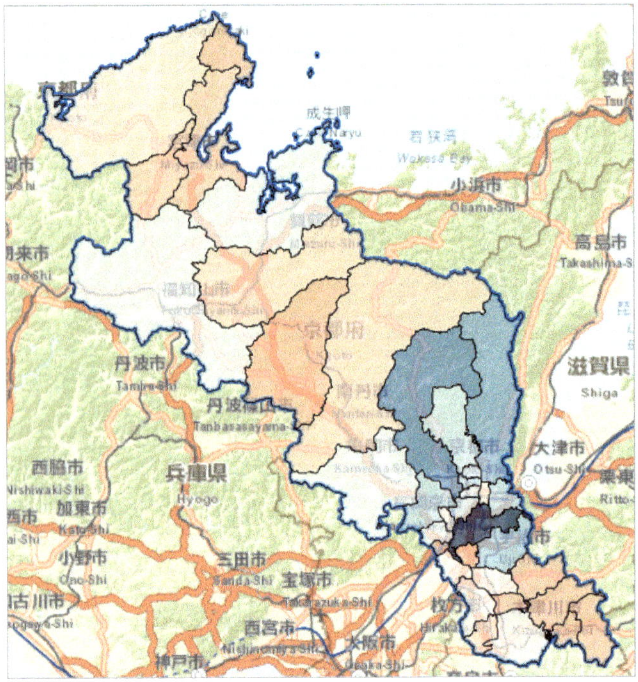


Now the household list values for each municipality are compared with the mean. The map and table use color-coding to represent where values are higher than or lower than the mean.

### Take it to the next level

You can explore the results to determine where projected household growth is higher than average and may be a good candidate for your new housing development.

Click the municipality with the highest projected household growth rate, and in the pop-up menu, click **Create site**. You can use this site as the area of interest for a new points of interest search and search for competitors. With the competitors mapped, you can identify whether the market is oversaturated, underserved, or competitive.

16. Change the **Data Source** back to **United States (Esri 2025)**.

# Tutorial 11-2: Run benchmark comparisons in Business Analyst Pro

In this tutorial, you will compare health-care spending variables for census tracts in Baldwin County, Alabama.

## Create a project

First, you'll need to set up your project. To run benchmark comparisons using Business Analyst Pro, you must have locations on the map to compare. For example, you can generate rings around points, create drive-time or walk-time areas, or draw geographic boundaries. Once these locations are saved to a layer in the project, you can compare them using the benchmark comparisons workflow.

1. Start ArcGIS Pro. If prompted, sign in using your licensed ArcGIS organizational account.
2. Under **New Project**, click **Map**.
3. In the **New Project** dialog box, apply the following settings:
    - **Name**: Ch 11 - Benchmark comparisons
    - **Location**: C:\Top20BusinessAnalyst

    **Tip:** This book uses **C:\Top20BusinessAnalyst** for the folder location, but you can use any folder.

4. On the ribbon, click the **Analysis** tab, and in the **Geoprocessing** group, click **Tools**.
5. In the **Geoprocessing** pane, click the **Toolboxes** tab.
6. Expand the **Business Analyst Tools** section. Expand the **Trade Areas** toolset and click **Generate Standard Geography Trade Areas**.

7. In the **Generate Standard Geography Trade Areas** tool, apply the following settings:

   - For **Geography Level**, select **Census Tracts (US.Tracts)**.
   - For **Output Feature Class**, type BaldwinCounty_TA.
   - For **Input Type**, select **List**.
   - For **Geography IDs List**, click the browse button. Select **US by Tracts** > **Alabama** > **Baldwin County**. Check the box for **Baldwin County** to select all the tracts in the county. Click **OK**.

8. At the bottom of the pane, click **Run**.

## Run benchmark comparisons

Now you can start running benchmark comparisons.

9. On the ribbon, click the **Analysis** tab. In the **Workflows** group, click **Business Analysis**.
10. At the bottom of the drop-down menu, make sure that the **Business Analyst Data Source** is set to **United States** using the most recent release of data, such as **Esri 2025**.
11. From the **General Workflows** section, click **Benchmark Comparisons**.
12. In the **Benchmark Comparisons Properties** pane, for **Input Features**, click **BaldwinCounty_TA**.

    You are using the standard geography census tracts that you previously added for your benchmark comparison.
13. For **Name Field**, make sure that **NAMES** is selected.
14. Next to **Variables**, click the **Add** button.
15. From the **Data Browser** window, search for and add the following variables:
    - **2025 Health Care (Avg)**
    - **2025 Pay Just About Anything for My Health: 4-Agree Completely (Index)**
    - **2025 Struggle with Health Care Costs: 4-Agree Completely (Index)**

    **Tip:** These variables are used in the **Health care spending** featured list for benchmark comparisons in Business Analyst Web App.
16. Click **OK** to close the **Data Browser** window.
17. In the **Benchmark Comparisons Properties** pane, click the **Comparison Methods** tab.
18. For **Statistics**, check the box for **Mean**.

19. For **Method**, click **Top and bottom**.

The Top and bottom comparison method highlights the top- and bottom-ranked values. It uses either Rank or Percentile, and you can define the Top and Bottom values.

20. At the bottom of the pane, click **Apply**.

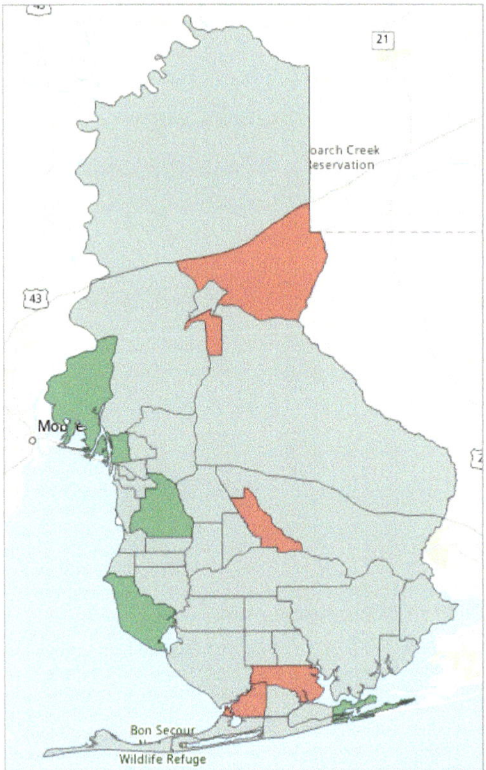


The results are displayed in the following places:

- The Contents pane has a new layer.
- The comparison layer appears on the map.
- The Results pane displays the results in summary, histogram, scatterplot, and table views.

### Take it to the next level

Open the results from the **Benchmark Comparison Properties** pane by clicking **View Results**. Click **Table** to view the results in a table. View the variable and measurement values for each feature, as well as the feature's comparison indicator.

## Summary

In this chapter, you learned how to run benchmark comparisons. You analyzed project household growth in Kyoto, Japan, using Business Analyst Web App. You also explored health-care spending variables to compare census tracts in Baldwin County, Alabama, using Business Analyst Pro.

### Workflow

1. Open your project.
2. Set the area of interest.
3. Select your analysis criteria.
4. Set your comparison method.
5. Explore the results.

# CHAPTER 12

# Creating threshold areas

## Objectives

- Run a threshold area analysis to gain information on surrounding demographics.
- Create different types of threshold areas based on a variety of variables.

## Introduction

In ArcGIS Business Analyst, you can map service areas using threshold areas. Instead of creating a trade area with a known ring, such as a 10-minute walking time or a two-mile radius from your site, you can map the distance needed to capture a variable, such as a population of 100,000 people. Threshold areas commonly use total population or households in an area, but you can also enrich your analysis using other variables from the data browser.

# Tutorial 12-1: Create threshold areas in Business Analyst Web App

In this tutorial, you will perform a threshold area analysis to plan community engagement initiatives for a main library branch in New York City.

## Create a project

To get started, you need to make a new Business Analyst project.

1. In your browser, navigate to bao.arcgis.com.
2. Enter your ArcGIS credentials and click **Sign In**.
3. On the **Home** tab, click **New project**. Name your project Ch. 12 - Threshold areas and click **Create**. Open your project.

   **Tip:** If you are prompted to clear the map of previous data, click **Yes**.

## Create threshold areas

Now that you've made your project, you can create threshold areas.

4. On the ribbon, click the **Run analysis** tab and click **Threshold areas**.

   By default, the workflow uses a single point location.

5. Make sure that **Single location** is checked and click **Next**.
6. In the search field, type 476 5th Ave, New York, NY and select the result with a zip code of 10018.

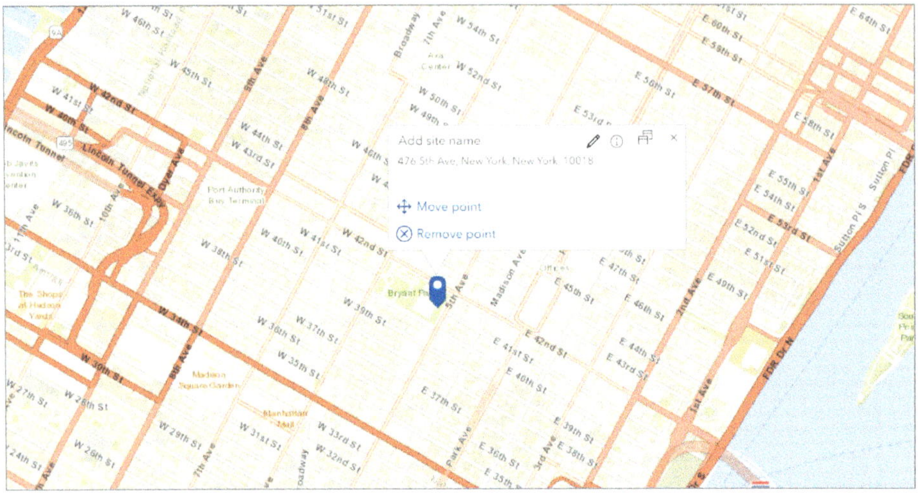


A pin appears on the map to mark the point.

7. Click **Next**.

   By default, the workflow uses the 2025 Total Population (Esri) variable with one threshold of 10,000 people. You'll add two more threshold amounts.

8. For **Threshold 2**, type 50000.

9. For **Threshold 3**, type 100000.

10. Click **Create areas**.

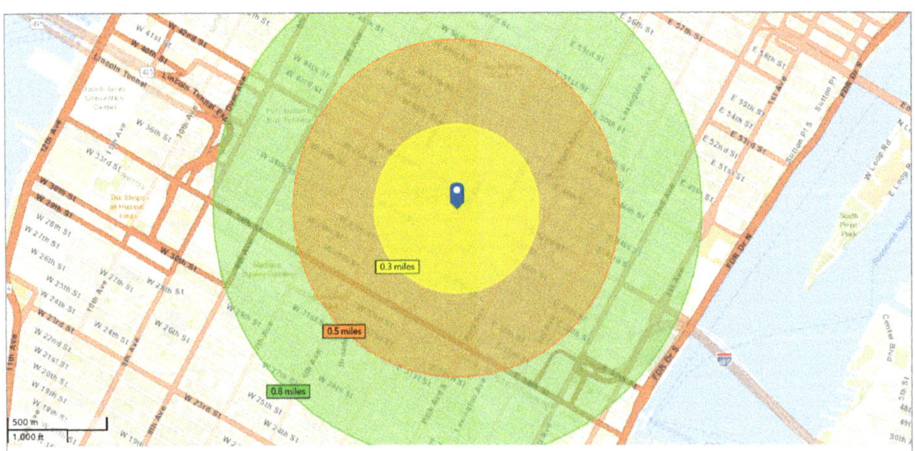


| Site name | Description | Threshold 1 | Buffer 1 (mi) | Threshold 2 | Buffer 2 (mi) | Threshold 3 | Buffer 3 (mi) |
|---|---|---|---|---|---|---|---|
| 476 5th Ave, New York, Ne... | 476 5th Ave, Ne... | 3,742 | 0.3 | 26,219 | 0.5 | 104,937 | 0.8 |

The results are displayed on the map and in the Threshold areas pane.

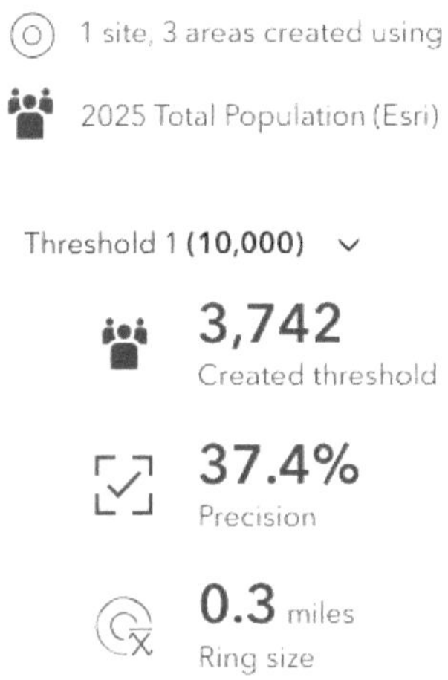


Threshold calculations are always necessarily an approximation—when the algorithm runs using real-world conditions, the output differs based on these conditions. There will be cases when the threshold values returned by the algorithm differ from the threshold values you set in the workflow. The values returned may be more or less than the requested threshold value based on:

- The underlying fabric of population density
- Urban structure such as high-rise development or suburban fabric
- Road layout, traffic patterns, driving speeds, and density
- The size, distribution, and actual placement of census block points

If the returned value is not what you expect, you can adjust the threshold settings slider in Business Analyst Web App to Best precision or manually adjust the values for Target percent difference and Increment size.

11. Click **Finish**. Close the **Threshold areas** workflow pane.

12. Repeat steps 4–7 to perform a new threshold analysis using the same address.

13. Under **Calculate using**, click the current variable and then click **Browse for a variable**.

14. In the **Data browser** window, search for and select **2023 Pop 3+ Enrolled in School (ACS 5-Yr)**.

15. Set the following thresholds:

- **Threshold 1**: 10000
- **Threshold 2**: 50000
- **Threshold 3**: 100000

In this case, the threshold measures the number of students in an area.

16. Click **Create areas**.

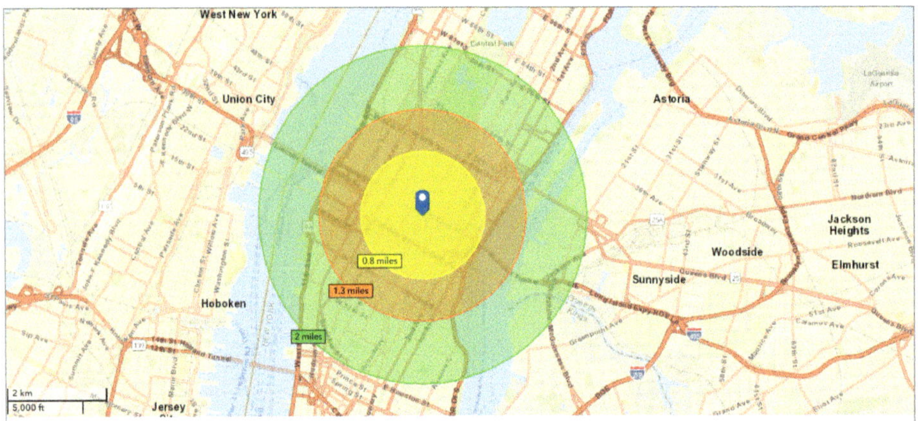


| Site name | Description | Threshold 1 | Buffer 1 (mi) | Threshold 2 | Buffer 2 (mi) | Threshold 3 | Buffer 3 (mi) |
|---|---|---|---|---|---|---|---|
| 476 5th Ave, New York, Ne... | 476 5th Ave, Ne... | 9,917 | 0.8 | 41,001 | 1.3 | 104,311 | 2 |

A new layer is created with threshold area rings. For this threshold area analysis, the rings are much larger using the school enrollment variable compared with the total population variable. The rings of the threshold area, or boundaries, change based on the density of the variable.

Analyzing the characteristics of your service area enables you to better understand the community and create strategies to more effectively serve community residents.

# Tutorial 12-2: Create threshold areas in Business Analyst Pro

In this tutorial, you will map Walmart store locations around Rochester, New York. To measure 50,000 customers in the area, you will create threshold areas based on five-minute drive times.

## Create a project

1. Start ArcGIS Pro and sign in using your licensed ArcGIS organizational account.
2. Under **New Project**, click **Map**.
3. In the **New Project** dialog box, for **Name**, type Ch 12 - Grocery store threshold areas. Save the project to **C:\Top20BusinessAnalyst**.

> **Tip:** This book uses **C:\Top20BusinessAnalyst** for the folder location, but you can use any folder.

## Perform a points of interest search

4. Zoom to the general area of Rochester, New York.

5. On the ribbon, click the **Analysis** tab.

6. In the **Workflows** group, open the **Business Analysis** drop-down menu. Under **General Workflows**, click **Points of Interest Search**.

   The Points of Interest Search pane appears.

7. For **Step One: Initial Search**, under **Name or keyword**, type Walmart and click **Next**.

8. At the bottom of the pane, click **Preview Map** to show all the Walmart stores that the search returned.

9. In the results list, expand **City**. Hover next to **Rochester** and click **Only**.
10. For **Output**, name the layer Walmart_Rochester.
11. Click **Finish**.

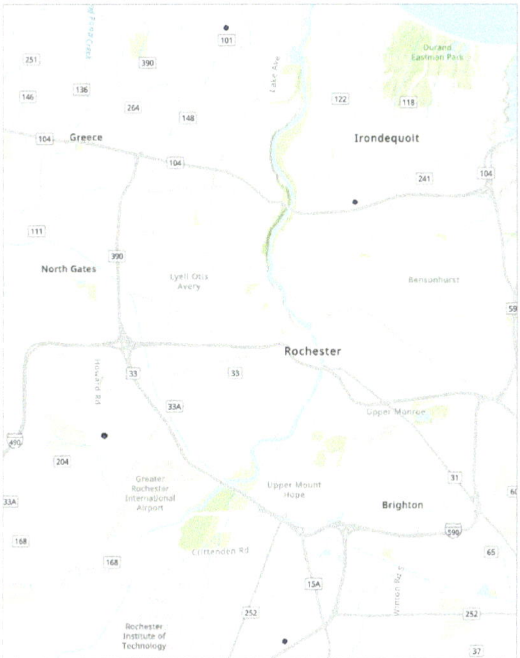


The store locations are saved to a new layer, visible on the map and in the Contents pane.

## Create threshold areas around stores

Now that points have been added to the map representing Walmart stores in Rochester, you can create threshold areas to determine a five-minute drive time based on the population of the area.

12. On the ribbon, click the **Analysis** tab. In the **Geoprocessing** group, click **Tools**.

13. In the **Geoprocessing** pane, search for Generate Threshold Drive Times and open the tool.

14. In the **Generate Threshold Drive Times** tool, apply the following settings:

    - For **Input Features**, select **Walmart_Rochester**.
    - For **Output Feature Class**, change the name to Walmart_Rochester_thresholds.
    - For **Threshold Variable**, click the **Add** button. From the **Data Browser** window, add the **2025 Total Population (Esri)** variable.
    - For **Input Method**, leave the default, **Values**.
    - For **Threshold Values**, type 50000.
    - For **Distance Type**, select **Driving Time**.
    - For **Distance Units**, leave the default, **Minutes**.
    - For **ID Field**, select **Esri PID**.

15. Click **Run**.

**Tip:** If prompted to estimate credits, at the top of the tool, click **estimate credits**. The tool is estimated to consume about 3 credits.

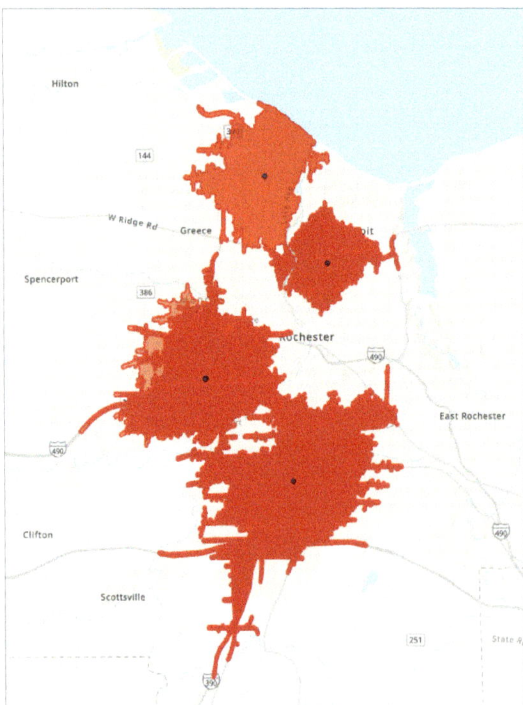

The threshold areas appear on the map. The layer's attribute table contains new fields that specify the variable value, actual value, radius length, and radius units.

## Summary

In this chapter, you created threshold areas using Business Analyst Web App and Business Analyst Pro. With the web app, you analyzed total population compared with student population to determine a library's service areas in New York City. With Business Analyst Pro, you mapped Walmarts around Rochester and then created threshold areas to determine how far customers would have to drive to reach your store.

### Workflow

1. Create a new project.
2. Choose a location or create sites based on points of interest.
3. Create threshold areas based on a variable.

# CHAPTER 13
# Performing void analysis

Coauthor Elif Bulut

## Objectives

- Compare an analysis area and a reference area to perform a void analysis.
- Use the void analysis to determine areas of need for senior care centers.
- Review senior care center information and compare against population demographics.

## Introduction

Void analysis is a tool that allows organizations and policymakers to identify gaps in services or resources in specific geographic areas.

The void analysis workflow analyzes an area you choose (the analysis area) to detect voids and gaps in specific businesses and services compared with another area (the reference area). A void means that a business or business type exists in the reference area but not in the analysis area, and a gap means that there are fewer of these businesses in the analysis area than in the reference area. Identify the area you want to analyze, select the business categories you want to evaluate, and view analysis results that break down voids and gaps in the selected businesses.

In practical terms, void analysis reveals what exists in one area but not in another, pinpointing which services are underrepresented.

> *Important: This chapter requires a Business Analyst Web App Advanced license.*

# Tutorial 13-1: Use void analysis to locate underserved senior citizens

In this tutorial, you will use Business Analyst Web App to identify gaps in senior care facilities in New Mexico between the analysis area (Sierra County) and reference area (Catron County). These insights will help pinpoint specific service gaps and highlight the level of unmet need.

## Create a project

To get started, you need to create a new Business Analyst Web App project.

1. In your browser, navigate to bao.arcgis.com.
2. Enter your ArcGIS credentials and click **Sign In**.
3. On the **Home** tab, click **New project**. Name your project Ch. 13 - Void analysis and click **Create**. Open your project.

   **Tip:** If you are prompted to clear the map of previous data, click **Yes**.

## Set up your map

To use the void analysis workflow, you'll first create sites for your analysis and reference areas.

4. On the ribbon, click **Define areas** and click **Select geography**.
5. Click **Select from full list**.
6. Scroll down the list of geographies and click **Counties**.
7. In the **Select State** list, click **New Mexico**.

   All the counties in New Mexico are listed for you to select from.
8. Check the boxes for **Catron County, NM** and **Sierra County, NM**.

9. Click **Finish**.

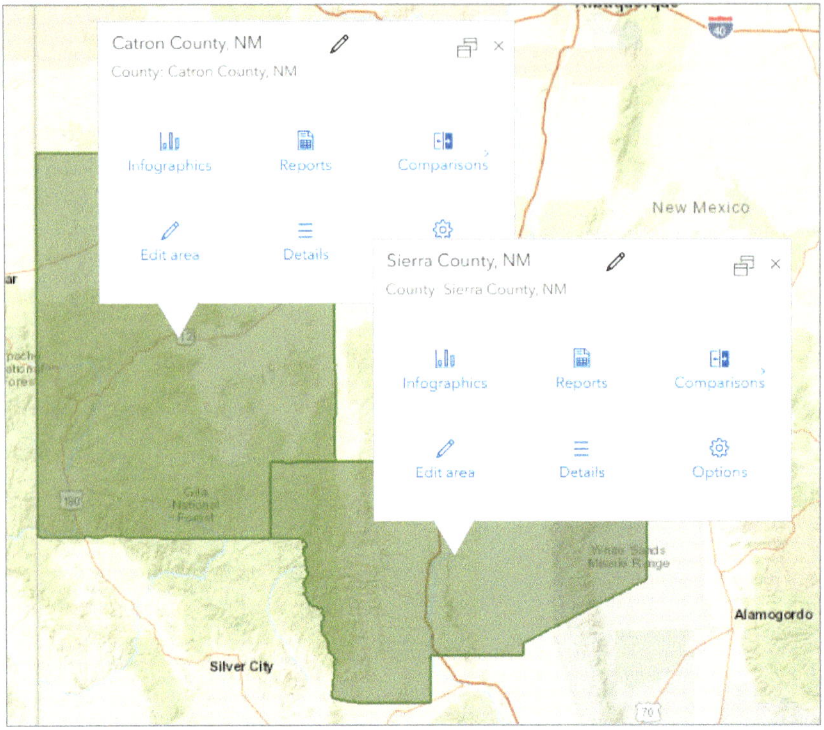


Both counties are added to the map as sites.

10. Click the **Close workflow** button to close the **Select geography** pane.

## Perform a void analysis

Now that you've made your sites, you can perform a void analysis to compare the number of senior care facilities in each county.

11. On the ribbon, click **Run analysis** and click **Void analysis**.

12. Click **Select analysis area** and check the box for **Sierra County, NM**. Click **Apply**.

    The analysis area is the area you select to detect voids and gaps in specific businesses and services.

13. Click **Select reference area** (or **Change reference area** > **Select area manually**) and check the box for **Catron County**. Click **Apply**.

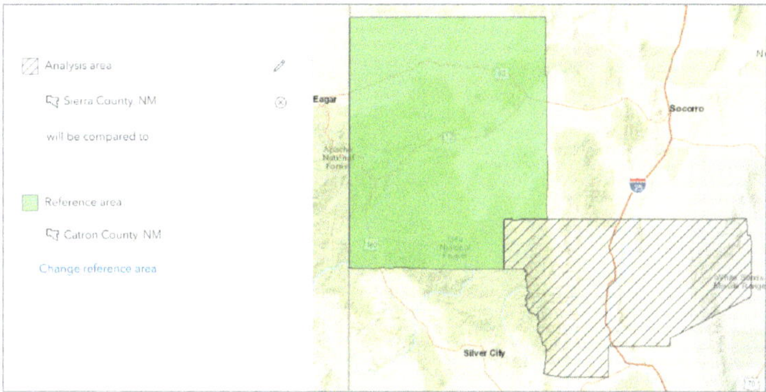


The reference area is what you are using to compare your selected analysis area to.

14. Click **Next**.

    Now you can select the business types (otherwise known as points of interest) that you want to analyze.

15. Click **Create custom list**.
16. Click **Enter keyword, category, etc. and** click the **Category** tab.
17. In the **Enter category name** search field, type senior care.
18. From the results, check the boxes for the following 14 categories:

    - **Senior Citizens Service Organizations**
    - **Senior Citizens Service**
    - **Elderly & Disabled Person Services**
    - **Adult Care Facilities**
    - **Retirement & Life Care Cmnty/Homes info**
    - **Life Care Communities**
    - **Senior Citizens Housing**
    - **Long Term Care Facility**
    - **Nursing Home Services**
    - **Residential Care Homes**
    - **Community Care Facilities-Adult Care**
    - **Day Care Centers-Adult**
    - **Adult Congregate Living Facilities**
    - **Retirement Communities & Homes**

    **Tip:** Sort the list by alphabetical order.

19. Click **Done**.

20. Click **Save list**. In the **Save list** window, for **List name**, type Senior care centers. Click **Save and run analysis**.

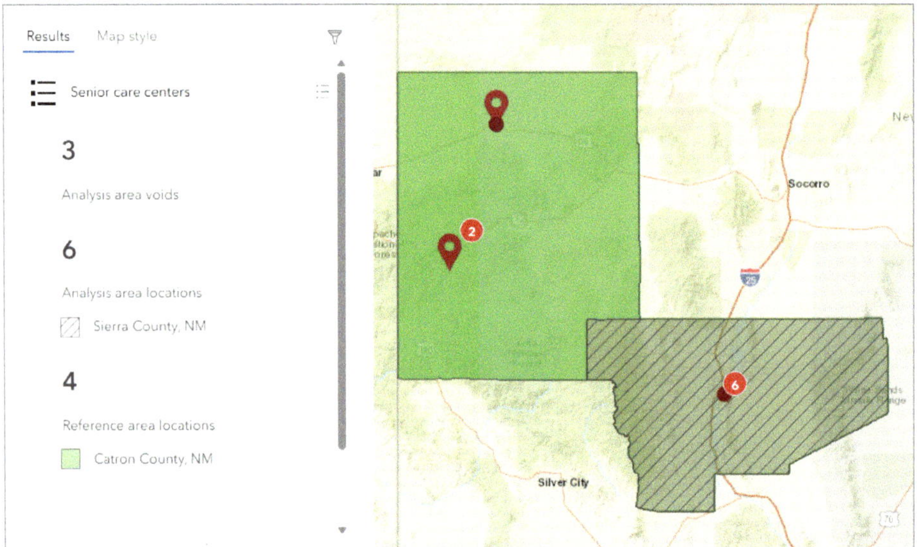


The results appear on the map and in the Void analysis pane. The results show that three voids were identified in Sierra County. These are senior care facilities that exist in Catron County (your reference area) but are missing in Sierra County (your analysis area).

21. Click **Full results.**

Now you can see the tabulated results.

22. Click the **Settings** button and check the box for **Show categories only**. Click **Done**. Turn on the **Advanced view** toggle button.

The advanced view provides more details about the analysis results and has automatically normalized the results based on population. Rows highlighted in red represent voids, meaning that the senior care facility exists in the reference area (Catron County) but not in the analysis area (Sierra County). Rows highlighted in blue represent senior care facilities that exist in both counties.

23. Next to **Normalize results by**, click the list and click **Browse for a variable**.

You will normalize the results by senior population in each county.

24. In the data browser, search for senior population.

25. Click the most recent senior population variable by Esri, such as **2025 Senior Population (Age 65+) (Esri).**

The advanced view of the void analysis results table now displays the senior populations of both the analysis area (Sierra County) and the reference area (Catron County), as well as the difference in senior population between them.

26. When you're done with your analysis, you can click **Export** to export your results to an Excel spreadsheet or click the **Close workflow** button to close the **Void analysis** pane.

    Your results cannot be saved as a map layer in the project pane.

## Summary

In this chapter, you created sites for two counties and used them as the analysis and reference areas in a void analysis. To perform your void analysis, you created a custom list of senior care facilities. You modified the settings for the full results to view the results normalized by the senior population. The void analysis revealed a significant gap in senior care services in Sierra County compared with Catron County.

### Workflow

1. Create your project.
2. Select two standard geographies to create your analysis and reference area sites.
3. Perform a void analysis.
4. Modify the settings for the full results table.

# CHAPTER 14

# Using trade area tools to measure overlap

## Objectives

- Create store locations using the points of interest search workflow.
- Generate store trade areas.
- Create a cannibalization report.

## Introduction

Business Analyst Pro has been designed to analyze the impact of site selection and business optimization strategies. When a company wants to expand its business, it needs to understand how much new business will be gained and how the new location may cannibalize or take away business from existing stores.

Cannibalization can also be used to understand the impact of nearby competitors, new businesses that open within a location's trade area, or the effect of a store closing and customers then needing to shop at a new or different location.

In Business Analyst Pro, you can use existing trade areas around sites to measure the amount of overlap between two or more trade area polygons. The overlap is measured in several ways, considering:

- The total overlap of all the trade areas, which is a measure of maximal cannibalization.
- The extent of overlap between any two polygons and the amount of overlap between each of the polygons. The proportion of overlap is typically unique to each polygon because the size of each trade area will vary, and therefore the total area of overlap will be different as a percentage of the polygon's area.

The Measure Cannibalization tool creates a new feature layer containing only the overlapping areas across polygons with detailed measurements of the overlap between each trade area. It also provides a comparison of the impact of the overlap on each overlapping polygon.

In this chapter, you'll use the Measure Cannibalization tool to explore how existing grocery stores' trade areas overlap. The cannibalization report can also be used to find the unique trade area for each store and analyze how stores compete.

## Tutorial 14-1: Create a small set of neighborhood grocery stores

You have six existing grocery stores in Redlands and Highland, California. Four locations are in suburban areas away from the freeway, serving residents who will drive up to seven minutes to do their grocery shopping. Each of the stores is merchandized to appeal to the unique demographics of these communities. Two locations are near a freeway and are predominantly aimed at commuters who want a convenient location to shop during the journey home or while visiting other stores in the area.

In this tutorial, you'll use the points of interest search workflow to add the grocery stores to the map and use the Generate Drive Time Trade Areas tool to create trade areas for each store.

### Create a project

1. Start ArcGIS Pro. If prompted, sign in using your licensed ArcGIS organizational account.
2. Under **New Project**, click **Map**.

3. In the **New Project** dialog box, apply the following settings:

   - **Name**: Ch 14 - Measure Cannibalization
   - **Location**: C:\Top20BusinessAnalyst

   **Tip:** This book uses **C:\Top20BusinessAnalyst** for the folder location, but you can use any folder.

4. Click **OK** to create your project.

## Add your grocery stores to the map

You have existing grocery stores in the Redlands market. To start your analysis, you'll zoom to Redlands, California, and create sites for each location.

5. On the ribbon, on the **Map** tab, find the **Inquiry** group. Click the **Locate** tool to open the **Locate** pane.

6. In the **Locate** pane, click in the search box and type Redlands, CA.

7. Select the first option, **Redlands, CA, USA**, under **ArcGIS World Geocoding Service**.

8. Make sure that option **A** is selected.

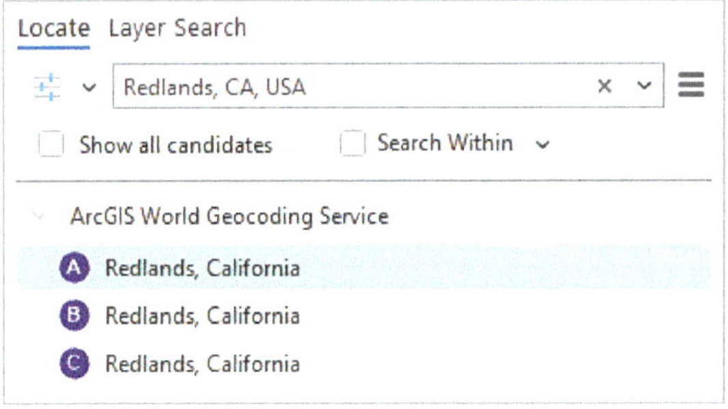


9. Close the **Locate** pane using the X button once the map has zoomed to Redlands.

10. On the ribbon, click the **Analysis** tab. In the **Workflows** group, click **Business Analysis** to open the gallery, and click **Points of Interest Search**.

    The points of interest search workflow allows you to create sites on your map. You will search for a local Southern Californian grocery store chain called Stater Bros.

11. In the **Points of Interest Search** pane, for **Step One: Initial Search**, click **Name or keyword**. In the search box, type Stater Bros and press Enter.

12. For **Step Two: Filter**, expand the **City** section. Uncheck the box for **City**. Check the boxes for **Redlands** and **Highland**.

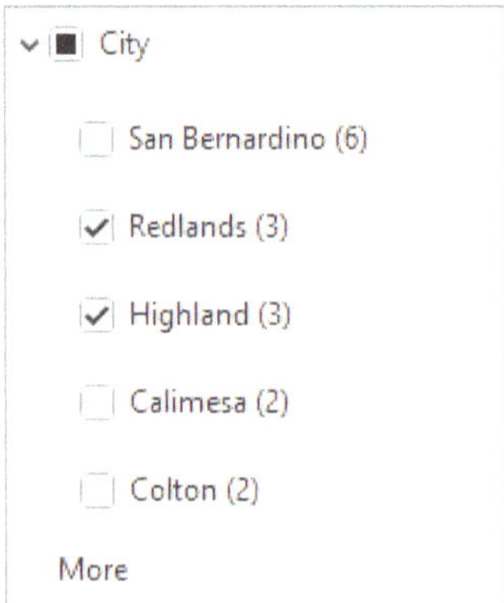


13. At the bottom of the tool, click **Preview Map** to see the Stater Bros. stores in Redlands and Highland on the map.

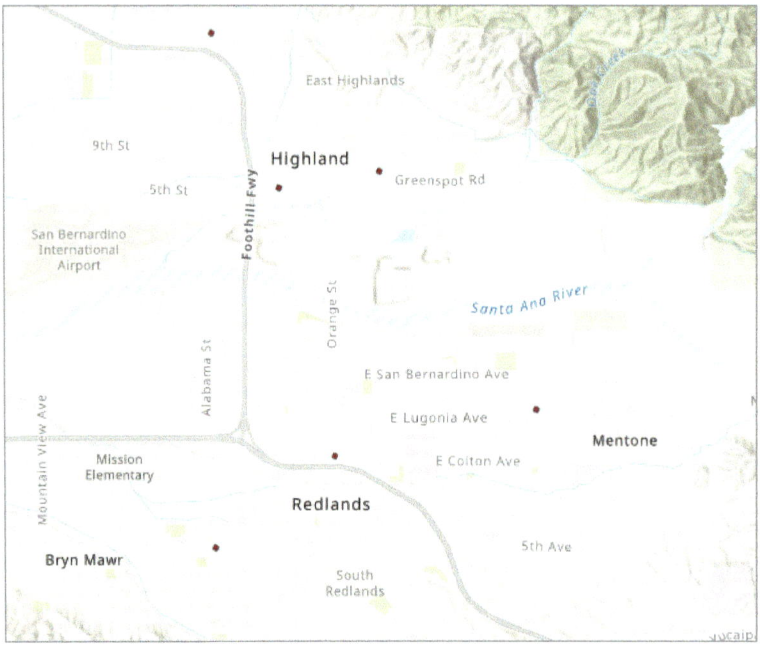


There should be six store location sites on the map.

14. Set the **Output** to Stater_Bros.

15. Click **Finish** to save the six stores.

    The points of interest are added to the map. Your map now has the six Stater Bros. grocery stores in it.

## Create trade areas around the stores

Now that you've added your stores, you need to create trade areas around each store. It is assumed that the drive-time trade area for each of the Stater Bros. stores is seven minutes, which caters to each distinct neighborhood market. You will use the Generate Drive Time Trade Areas tool to create the trade areas.

16. On the **Analysis** tab, in the **Geoprocessing** group, click **Tools**.

17. On the **Toolboxes** tab, expand the **Business Analyst Tools** > **Trade Areas** toolset. Click the **Generate Drive Time Trade Areas** tool.

18. In the **Generate Drive Time Trade areas** tool, apply the following settings:

    - For **Input Features**, select **Stater_Bros**.
    - For **Distances**, type 7.
    - For **ID Field**, select **Street**.

19. At the top of the tool pane, click **estimate credits**.

    This tool is estimated to consume about 3 ArcGIS credits.

20. Click **Run**.

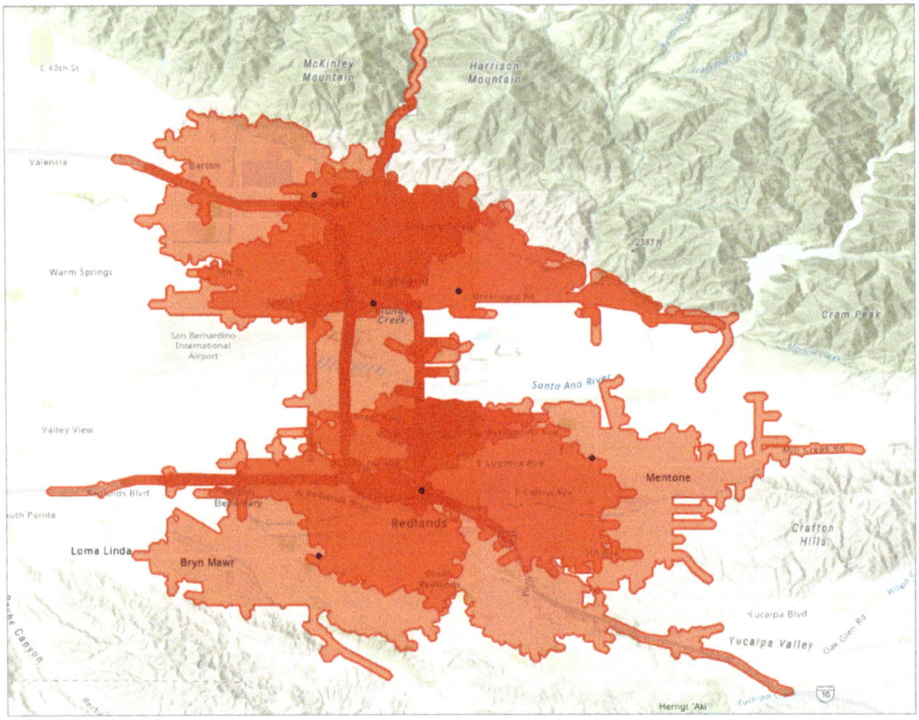


21. Now that you've created the trade areas for each store, you can visually see how they overlap.

## Tutorial 14-2: Measure cannibalization around each store and generate a report that explains the overlap

In this tutorial, you will turn this visual information into a quantitative report and feature layer that allows you to get information on each of the trade area interactions to see which location has the most impact on the other grocery stores. You will use the Measure Cannibalization tool to do this.

### Run the Measure Cannibalization tool

1. Click the back button in the **Geoprocessing** pane to return to the **Toolboxes** tab.
2. In the **Business Analyst Tools** > **Trade Areas** toolset, click the **Measure Cannibalization** tool.

3. In the **Measure Cannibalization** tool, apply the following settings:

   - For **Input Features**, select **Stater_Bros_GenerateDriveTimeTradeArea**.
   - For **Trade Area ID**, select **AREA_ID**.
   - For **Trade Area Description**, select **Street**.
   - For **Store ID Field**, select **AREA_ID**.
   - Check the box for **Create Report**.

4. Expand the **Report Options** section. For the **Output Report Folder**, click the browse button. In **Project** > **Folders**, for **Name**, type MeasureCannibalization.

   You can use the Additional Metrics option of the Measure Cannibalization tool to add demographic attributes that might help further quantify the cannibalization or measure the impact on market potential. Variables such as Food at Home can be used to estimate the spending on groceries and how much spending is shared or transferred between stores compared with the total spending on food in the combined trade area.

5. Expand the **Additional Metrics** section. Click the **Variables** add button. In the **Data Browser** window, search for and select the following metrics:

   - **2025 Food at Home**
   - **2025 Total Population (Esri)**
   - **2025 Total Households (Esri)**

6. Click **Run**.

   You can see on the map a set of new polygons, which shows the overlap and cannibalization for each of the stores.

7. On the map, click one of the cannibalization polygons to view the pop-up.

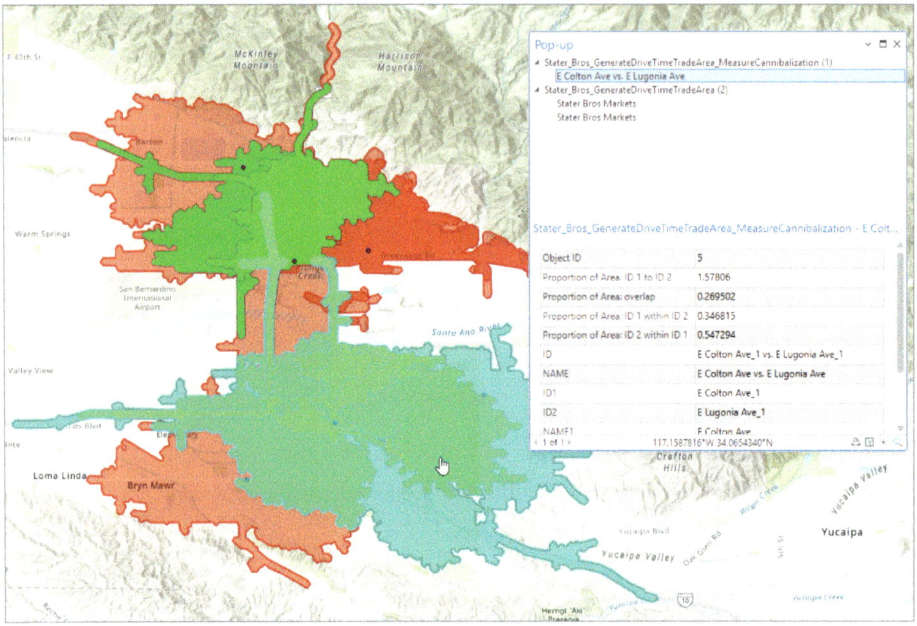


When you click on each area, you can see the cannibalization metrics for each area, including how the areas overlap with each other, the ratio of the overlap, and the names of each store that are impacted by the overlapping cannibalization.

## Review the cannibalization report

8. In File Explorer, go to C:**Top20BusinessAnalyst\Ch 14 - Measure Cannibalization\MeasureCannibalization**. Open **Measure Cannibalization .pdf**.

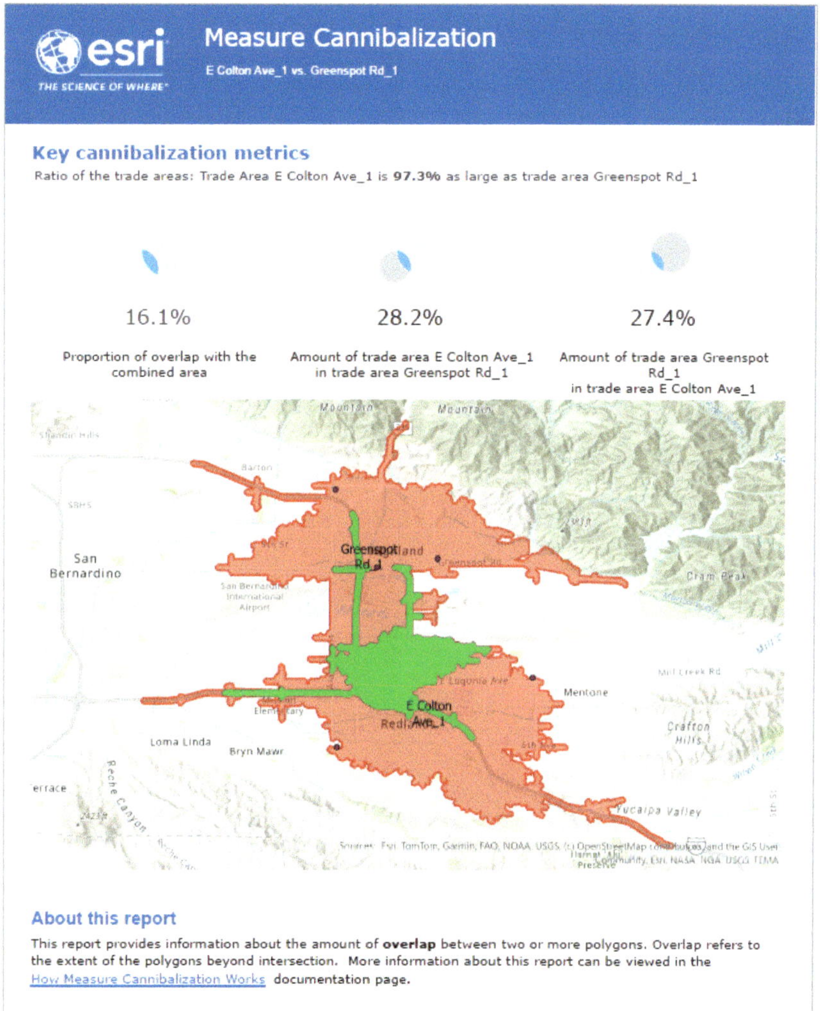


The report provides a boardroom-ready summary of the map, formatted with one page per store, containing clear information on the impact of the trade area overlap. With this information, you can devise a strategy to move, close, or open new stores; compare the effects of cannibalization for each store; or investigate what factors influence sales by understanding the demographics further.

In the preceding example, the Colton Ave. and Greenspot Rd. stores share 16.1% of the combined trade area of the two stores. The trade area of the Colton Ave. store that is in the Greenspot Rd. store trade area is 28.2%, whereas the reverse overlap is 27.4%. The report also explains that the Colton Ave. trade area is 97.3% as large as the Greenspot Rd. store.

9. In the report, scroll down to see the metrics for each area.

Using additional metrics rather than just raw area helps you make better decisions about what the impact of cannibalization truly is, compared with using the size of the overlapping areas alone. Locations with denser populations or higher spending on food at home may highlight different patterns and proportions in spending or population, compared with just the overlap in area.

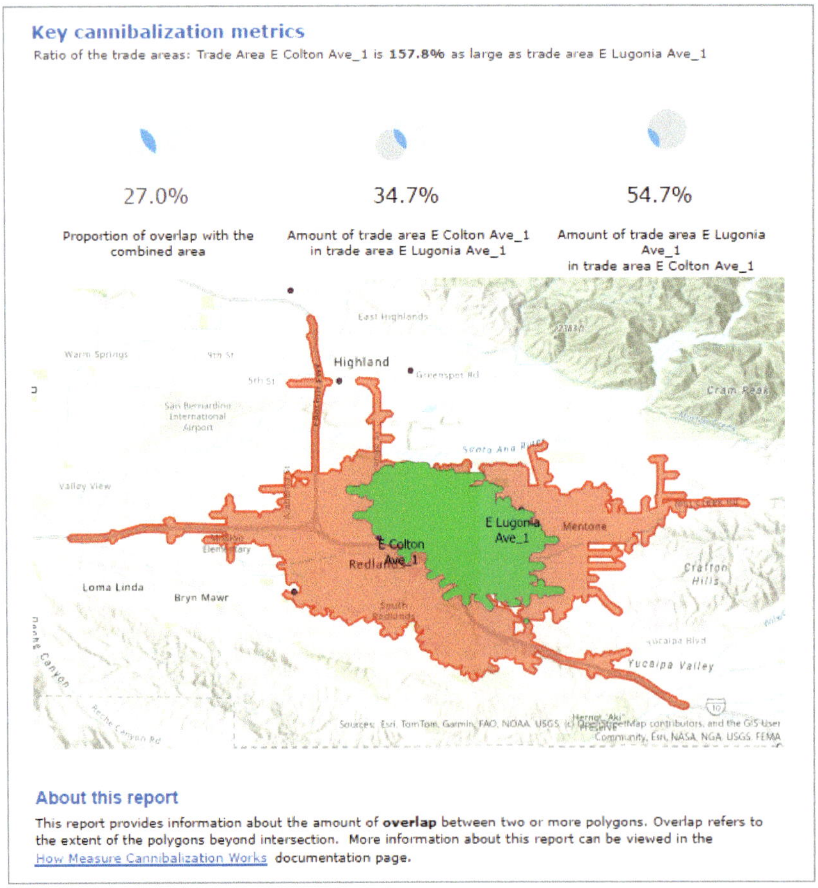


In the preceding example, the overlapping area (measured in square miles) between the Colton Ave. and Lugonia Ave. stores is 27.0%. However, when you look at the 2025 Food at Home metric, the cannibalization statistic is 40.1%, representing $78.8 million in a combined market of $196.7 million.

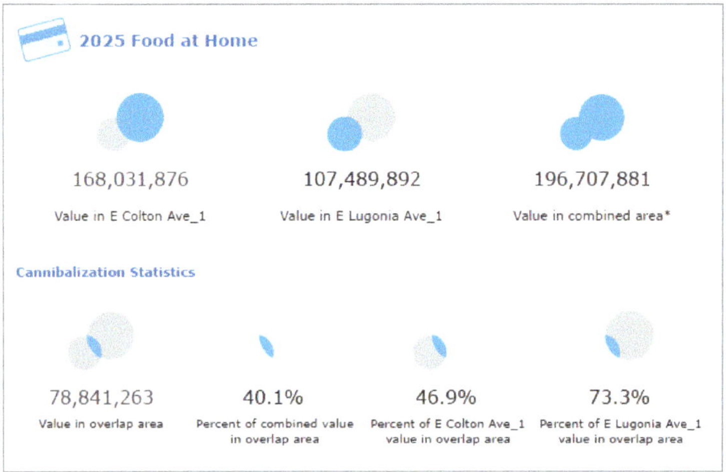


The Colton Ave. store may be performing both an offensive and a defensive strategy for Stater Bros. It is near the freeway and downtown Redlands, creating an opportunity to gain casual traffic on the freeway or consumers visiting downtown, while also working to offer two stores within the $78 million food at home market. The Lugonia Ave. store has an exclusive trade area, and having an overlap with the Colton Ave. store may mean that Stater Bros. is able to cover more of the market and outperform the competition in this market.

## Summary

In this chapter, you used Business Analyst Pro to add places to the map using the points of interest search workflow and then created trade areas around each site. You also applied the Measure Cannibalization tool to create statistics and a report about how the trade areas of the stores overlap.

### Workflow

1. Create your project.
2. Add your stores.
3. Create trade areas around each store or point of interest.
4. Perform cannibalization analysis to understand how each location overlaps and impacts the other stores.
5. Present the results in a map and in a ready-to-use report.

# CHAPTER 15

# Working with territory design

## Objectives

- Set up territories using standard geographies.
- Balance territories using a demographic variable.
- Solve territories to create a map of balanced territories.

## Introduction

Territory design sits at the intersection of GIS analysis and field operations. A territory is a geographic area used for assigning work, tasks, and management in physical space. Territory design is based on the physical space and factors in geographic information (such as facility locations, travel barriers, and proximate areas) as well as business information (such as work capacity, personal preferences, and historical decisions).

# Tutorial 15-1: Design territories in Business Analyst Pro

In this tutorial, you'll learn the fundamentals of territory design by creating a map for election workers in Iowa.

## Create a project

1. Start ArcGIS Pro. If prompted, sign in using your licensed ArcGIS organizational account.

2. Under **New Project**, click **Map**.

3. In the **New Project** dialog box, apply the following settings:

   - **Name**: Ch 15 - Iowa election territories
   - **Location**: C:\Top20BusinessAnalyst

   **Tip:** This book uses **C:\Top20BusinessAnalyst** for the folder location, but you can use any folder.

4. Click **OK** to create your project.

## Add locations to the map

5. On the ribbon, click the **Analysis** tab. In the **Geoprocessing** group, click **Tools**.

6. In the **Geoprocessing** pane, search for Generate Standard Geography Trade Areas and open the tool.

   Standard geography trade areas are administrative and political boundaries, such as states, counties, and zip codes.

7. In the **Generate Standard Geography Trade Areas** tool, apply the following settings:

   - For **Geography Level**, select **Counties (US.Counties)**.
   - For **Output Feature Class**, type Iowa_Counties.
   - For **Input Type**, select **List**.
   - Next to **Geography IDs List**, click the browse button. Select **US by Counties** and then select **Iowa**. Check the box for **Iowa** to select all 99 counties in the state and click **OK**.

8. At the bottom of the pane, click **Run**.

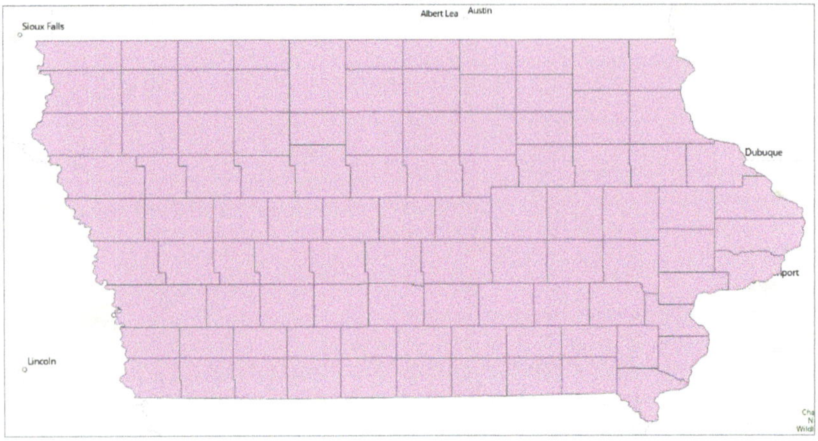


The tool runs, and a layer containing all the counties in Iowa is added to the map.

## Enrich the layer

9. In the **Geoprocessing** pane, click the back button.
10. In the search bar, type Enrich Layer. In the results, click the **Enrich Layer** tool to open it.
11. In the **Enrich Layer** tool, apply the following settings:
    - For **Input Features**, select **Iowa_Counties**.
    - For **Output Feature Class**, type Enriched_Iowa_Counties.
    - Next to **Variables**, click the **Add** button. In the **Data Browser** window, search for Vote in National Elections. Check the box for **2025 Vote in National Elections: 3-Always** and then click **OK**.

    This variable provides data on the number of likely voters in each county, which will be useful for predicting the workload of election workers.
12. At the top of the tool, click **estimate credits**.

    The tool is estimated to consume about 0.99 credits.
13. Click **Run**.

    A new layer of Iowa counties is created, containing the voting behavior data you selected. These counties will form the base level of the territory solution.

## Create a territory solution layer

14. On the **Analysis** tab, in the **Workflows** group, open the **Business Analysis** menu. Under **General Workflows**, click **New Territory Design Solution**.

15. In the **Create Territory Solution** window, apply the following settings:

    - For **Input Features**, select **Enriched_Iowa_Counties**.
    - For **Territory Solution Name**, type Iowa Election Territories.
    - For **ID Field**, select **Object ID**.
    - For **Names Field**, select **NAMES**.
    - Leave the defaults for **Territory Level Name** and **Default Territory Name**.

16. Click **OK**.

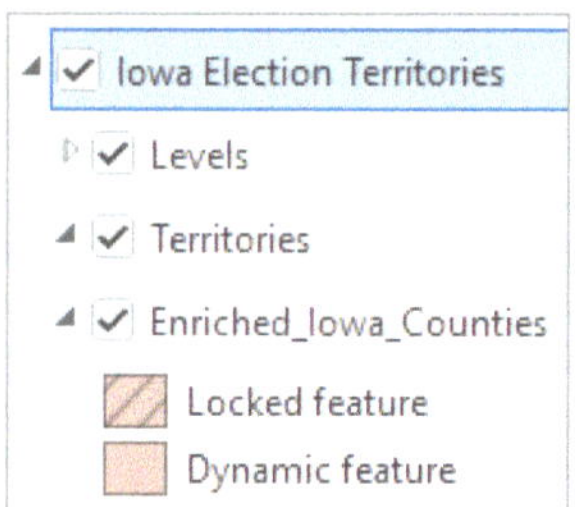


The empty territory solution layer is created, and you are ready to create territories.

## Add a variable to the analysis

17. In the **Contents** pane, click the **Iowa Election Territories** group layer to select it.

18. On the ribbon, click the **Territory Design** tab. In the **Analysis** group, click **Add Variables**.

19. In the **Add Level Variables** tool, apply the following settings:

    - For **Input Territory Solution**, select **Iowa Election Territories**.
    - For **Level**, select **Territories[1]**.
    - For **Base Level**, select **Enriched_Iowa_Counties[0]**.
    - Next to **Variables**, click the **Add many** button to open the list and select **2025 Vote in National Elections: 3-Always**.

20. Apply the other variables' settings as follows:

    - For **Statistic**, select **Sum**.
    - Leave **Field Name** as the default.
    - For **Field Alias Name**, type Likely Voters.

21. Click **OK**.

    The territory solution now contains the voting variable as a factor in territory creation.

## Balance the territories

22. On the **Territory Design** tab, in the **Analysis** group, click **Balance Variables**.

23. In the **Set Balance Variables** tool, apply the following settings:

    - **Input Territory Solution**: Iowa Election Territories
    - **Level**: Territories[1]

24. Under **Balance Variables**, apply the following settings:

    - **Variable**: Likely Voters
    - **Weight**: 100

25. Click **OK**.

    The territory solution is now set up to use the voting variable as the deciding factor in creating balanced territories.

## Solve the territories

26. On the **Territory Design** tab, in the **Analysis** group, click **Solve**.

27. In the **Solve Territories** pane, apply the following settings:

    - For **Input Territory Solution**, select **Iowa Election Territories**.
    - For **Level**, select **Territories[1]**.
    - For **Number of Territories Method**, select **User Defined**.
    - For **Number of Territories**, type 20.

28. Click **OK**.

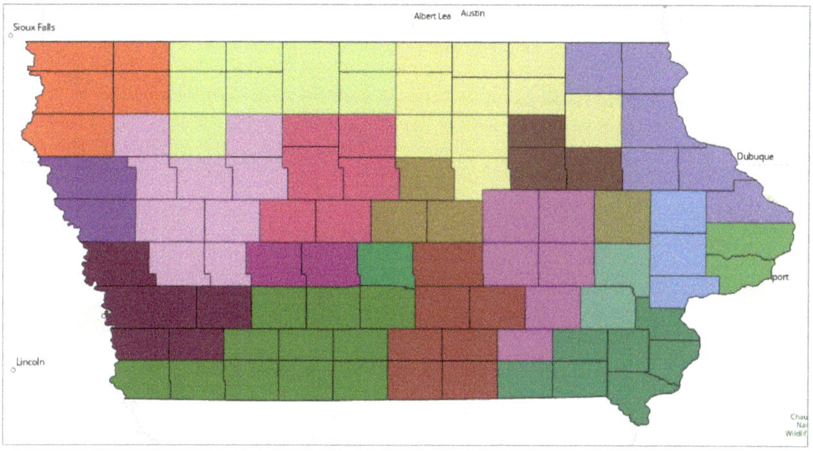


The map displays color-coded territories, which are also shown in the Contents pane.

Looking at the territories, a few things become apparent.

- Several single-county territories are concentrated around population centers (Des Moines, Cedar Rapids, Iowa City).
- Less-populated regions tend to be grouped together.
- A handful of awkwardly shaped territories could possibly be improved.

29. On the **Territory Design** tab, in the **Edit Territories** group, click **Modify Territories**.

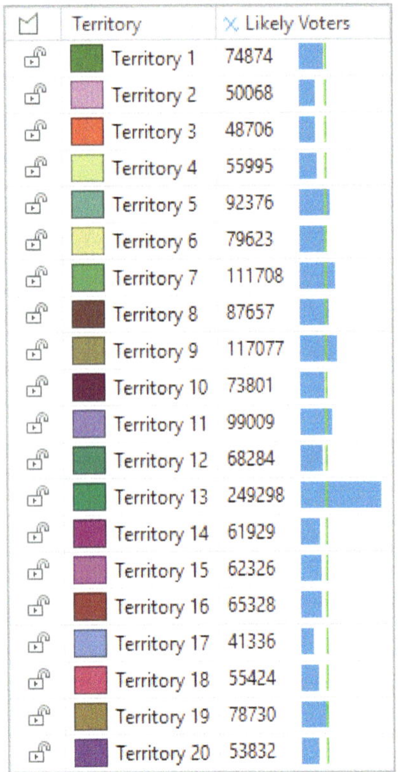

| Territory | Likely Voters |
|---|---|
| Territory 1 | 74874 |
| Territory 2 | 50068 |
| Territory 3 | 48706 |
| Territory 4 | 55995 |
| Territory 5 | 92376 |
| Territory 6 | 79623 |
| Territory 7 | 111708 |
| Territory 8 | 87657 |
| Territory 9 | 117077 |
| Territory 10 | 73801 |
| Territory 11 | 99009 |
| Territory 12 | 68284 |
| Territory 13 | 249298 |
| Territory 14 | 61929 |
| Territory 15 | 62326 |
| Territory 16 | 65328 |
| Territory 17 | 41336 |
| Territory 18 | 55424 |
| Territory 19 | 78730 |
| Territory 20 | 53832 |

Territory 13, which contains Des Moines, has a much larger number of likely voters than any other territory.

## Take it to the next level

You can use the **Modify Territories** pane to adjust the territories, such as switching their assignments or dividing the Des Moines territory into multiple territories.

# Summary

In this chapter, you used territory design in Business Analyst Pro to map and balance the workload for county-level election workers in Iowa.

## Workflow

1. Enrich a layer of base geographies using a demographic variable.
2. Create a territory solution layer.
3. Add the demographic variable to the analysis and use it to balance territories.
4. Solve the territories.

# CHAPTER 16

# Using the Huff model

## Coauthor Darren Cook

## Objectives

- Use purchasing data in a Business Analyst Pro project.
- Use the Huff model tool to predict customer behavior.
- Make a decision based on the model's output.

## Introduction

ArcGIS Business Analyst includes many tools to assist in decision-making, especially when it comes to choosing a location for a business or service facility. You might use Business Analyst Web App to perform a smart map search, as in chapter 9, that will identify the top locations for your enterprise and even rank them to help you pick the best one.

But how do you go from "here's my best guess based on local demographics" to modeling sales potential of the new location and gauging its impact on existing locations? How can you land on a solid confirmation of your choice, supported by business data?

One approach is to bring your analysis into Business Analyst Pro and use the Huff model to project demand and potential sales of your new location. This model allows you to attach specific probability to a candidate location. The model creates probability rates that indicate the likelihood that consumers will choose this new location over other existing locations. These rates are created for all areas of the market, defined by the extent of the Sales Potential features parameter in the tool. In this chapter, you will estimate the probability of customers visiting a coffee shop you are planning to open in Milwaukee, Wisconsin.

## Tutorial 16-1: Enrich block groups

In this tutorial, you will add Milwaukee County block groups and enrich them with consumer spending at coffee shops. You will then add coffee shop locations to your map and select a candidate location.

### Create a project

1. Start ArcGIS Pro. If prompted, sign in using your licensed ArcGIS organizational account.
2. Under **New Project**, click **Map**.
3. In the **New Project** dialog box, apply the following settings:
   - **Name**: Ch 16 - Huff model
   - **Location**: C:\Top20BusinessAnalyst

   **Tip:** This book uses **C:\Top20BusinessAnalyst** for the folder location, but you can use any folder.

4. Click **OK** to create your project.

### Add block groups to the map

5. On the ribbon, click the **Analysis** tab. In the **Geoprocessing** group, click **Tools**.
6. In the **Geoprocessing** pane, search for and open the **Generate Standard Geography Trade Areas** tool.

   Standard geography trade areas can be made up of any type of census, administrative, political, or other geography, including postal boundaries.

7. In the **Standard Geography Trade Areas** tool, apply the following settings:
   - For **Geography Level**, select **Block Groups (US.BlockGroups)**.
   - For **Output Feature Class**, type MilwaukeeBlockGroups.

- Leave **Input Type** as **List**.
- For **Geography IDs List**, click the browse button. In the **Select Geography** window, select **US by Block Groups** > **Wisconsin** > **Milwaukee County**. Check the box for **Milwaukee County**, selecting all 860 block groups in the county, and click **OK**.

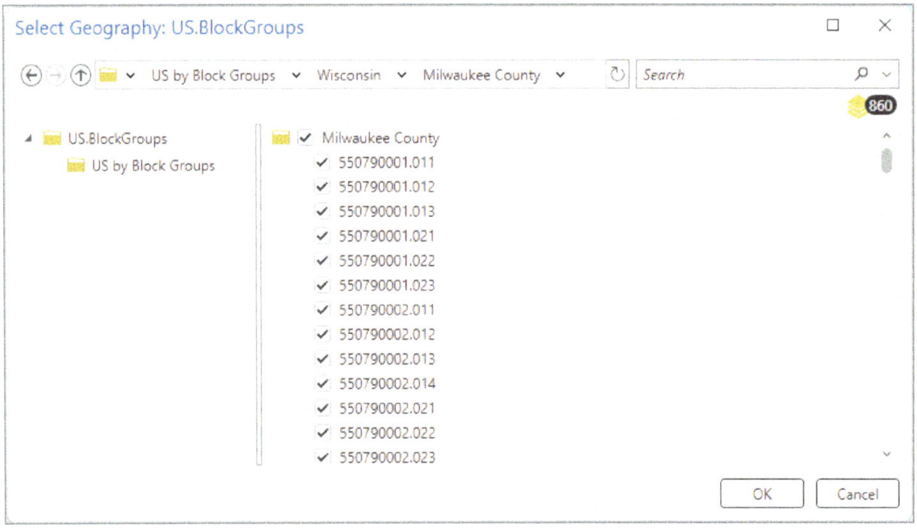


8. At the bottom of the tool pane, click **Run**.

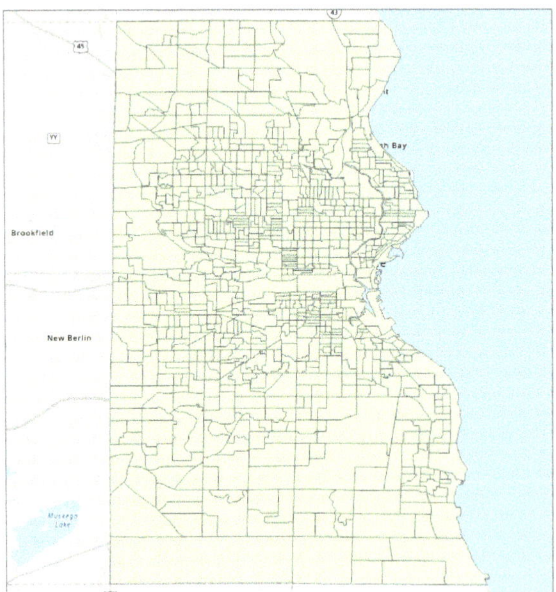


A layer of block groups in Milwaukee County is added to the map.

## Enrich block groups with coffee purchasing data

9. In the **Geoprocessing** pane, click the back button to return to the search.

10. Search for and open the **Enrich Layer** tool.

11. In the **Enrich Layer** tool, apply the following settings:

    - For **Input Features**, select **MilwaukeeBlockGroups**.
    - For **Output Feature Class**, type Milwaukee_BGs_coffee.
    - For **Variable**, click the **Add** button. In the **Data Browser** window, search for coffee. In the results, if necessary, expand **2025 Food at Home - Nonalcoholic Bev (Consumer Spending)**. Next to **2025 Coffee**, select **Avg** and deselect **#**. Check the box next to the variable and click **OK**.

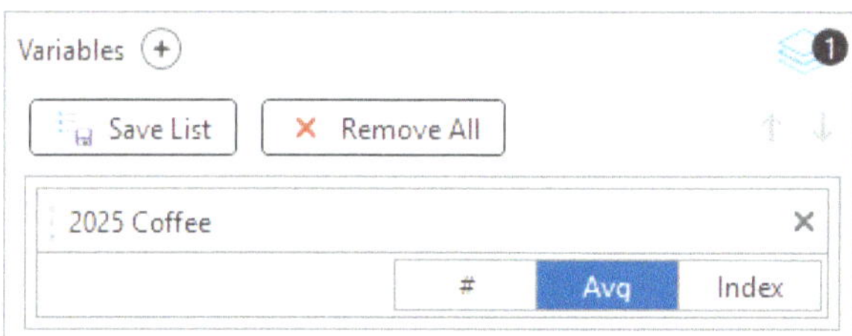


12. At the top of the pane, click **estimate credits**.

    The tool is estimated to consume about 8.6 credits.

13. Click **Run**.

    A layer of block groups in Milwaukee, enriched with coffee purchasing data, is added to the map.

14. In the **Contents** pane, right-click **MilwaukeeBlockGroups** and click **Remove** because you will not need it anymore.

## Add coffee shops to the map

15. On the ribbon, on the **Analysis** tab, locate the **Workflows** group. Click the **Business Analysis** gallery and click **Points of Interest Search**.

    The Points of Interest Search pane appears.

16. For **Step One: Initial Search**, under **Name or keyword**, type coffee and click **Next**.

17. For **Step Two: Filter**, expand the **City** section. Hover over **Milwaukee** and click **Only**.

18. Name the **Output** layer Milwaukee_coffeeshops. Click **Finish**.

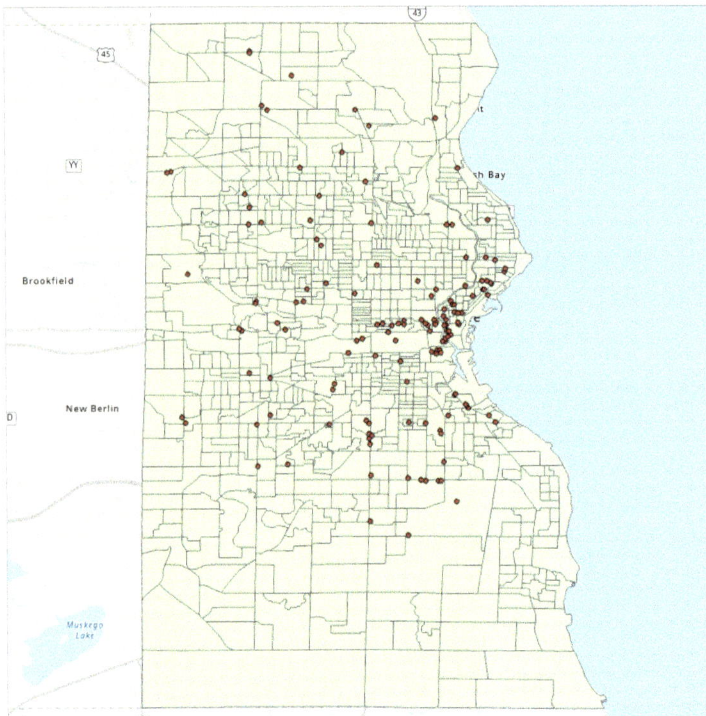


The map now contains a layer of coffee shops in Milwaukee.

## Create a fictional candidate location

19. In the **Geoprocessing** pane, click the back button to return to the search. Search for and open the **Create Feature Class** tool.

20. In the **Create Feature Class** tool, apply the following settings:

- For **Feature Class Location**, leave it as the default of your project's .gdb file.
- For **Feature Class Name**, type Candidate_location.
- For **Geometry Type**, select **Point**.
- For **Template Datasets**, select **Milwaukee_coffeeshops**.
- For **Coordinate System**, select **Current Map [Map]**.
- Leave the rest of the parameters as is.

21. Click **Run**.

    You have created an empty feature class where you can store a fictional point location representing the coffee shop you are planning to open in Milwaukee. You'll now add the fictional location to the feature class.

22. On the ribbon, click the **Edit** tab. In the **Features** group, click **Create**.
23. In the **Create Features** pane, click the **Candidate_location** layer.
24. On the map, place a point at a location of your choice.

    The image shown has a point placed near the water.

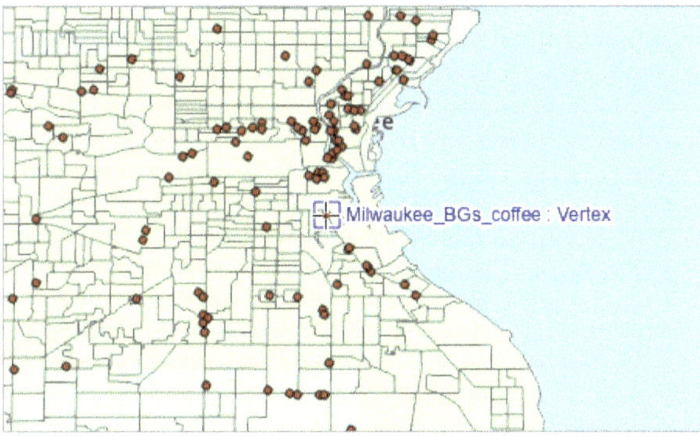


25. In the **Contents** pane, right-click the **Candidate_location** layer and click **Attribute Table**.

    Notice that, because you used the Milwaukee_coffeeshops layer as the template for this new layer, the fields are the same as that layer, but they are empty.

26. Scroll to the **Square Foot Maximum** column, double-click the cell, and type 2500. Press Enter.

    You'll use the square footage of the coffee shops to estimate attractiveness when running the Huff model. By adding the square footage of the candidate location to its attribute table, you are providing the data necessary to compare it with the other coffee shops.

27. For the **EsriPID** column, click the cell and enter a random value, such as X1. Press Enter.
28. Close the attribute table.
29. On the **Edit** tab, in the **Manage Edits** group, click **Save**.

# Tutorial 16-2: Create a Huff model

In this tutorial, you will use the coffee shops layer and your candidate coffee shop location to run the Huff model.

## Run the Huff model

30. On the ribbon, click the **Analysis** tab. In the **Workflows** group, click the **Business Analysis** gallery and click **Run Huff Model**.

31. In the **Huff Model** tool, apply the following settings:

    - For **Input Facility Features**, select **Milwaukee_coffeeshops**.
    - For **Facility ID Field**, select **Esri PID**.
    - For **Input Candidate Features**, select **Candidate_location**.
    - For **Candidate ID Field**, select **Esri PID**.
    - For **Input Sales Potential Features**, select **Milwaukee_BGs_coffee**.
    - For **Sales Potential ID Field**, select **IDField**.
    - For **Sales Potential Field**, select **2025 Coffee: Average**.
    - Under **Attractiveness Variables**, for **Existing Facilities Value**, select **MAX_SQFT**.
    - For **Candidates Location Value**, select **MAX_SQFT**.
    - Leave the **Exponent** value as **1**.
    - Leave the **Distance Exponent** value, as well as the other parameters, as their defaults.

32. Click **Run**.

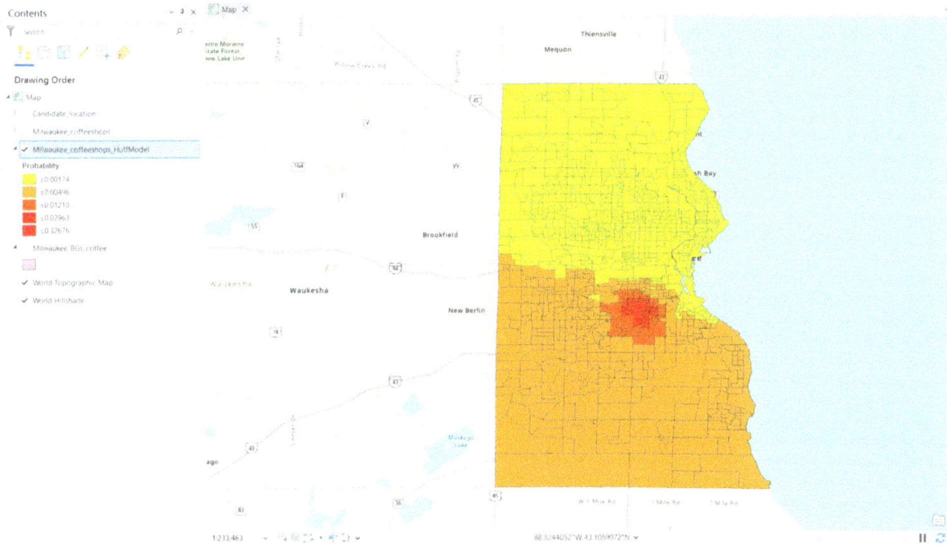


The tool creates a layer that symbolizes the probability that customers in each respective block group will visit the new coffee shop location. Areas mapped in yellow are block groups with the lowest probability and areas mapped in red represent the highest probability.

## Summary

In this chapter, you used the Huff model to predict the likelihood of customers visiting your new coffee shop based on the following factors: coffee purchasing behavior, size of the store, and distance from the store.

### Workflow

1. Enrich block groups with coffee purchasing data.
2. Map coffee shops in the area.
3. Create a candidate location for your new coffee shop.
4. Run the Huff model to map the probability of consumers visiting and purchasing at the site.

# CHAPTER 17

# Performing distance analysis using customer data

## Objectives

- Use customer and store data to create a customer-derived trade area.
- Create different types of trade areas based on data type.
- Analyze the relationship between trade areas.

## Introduction

A customer-derived trade area (CDTA) is an enclosed shape, technically a convex hull, around a proportion of customers based on either their distribution or an attribute weight, such as sales, number of visits, or other condition of interest. For example, retailers use CDTAs to identify a trade area based on customers or detected store visits. Health-care organizations can create CDTAs to map areas with patients who might need specialized care, based on the demand for pediatric or geriatric medicine. Warehousing operations use CDTAs to identify market areas based on demand for goods or services, using spending or potential.

Business Analyst Pro includes two key tools for creating trade areas based on customer data: the Assign Customers by Distance tool and the Generate Customer Derived Trade Areas tool. To create customer-derived trade areas, you must first assign each customer to a store.

In this chapter, you'll learn how to link customers with the stores they visit and then draw a trade area for each store based on its customers.

# Tutorial 17-1: Map and assign customers to stores

First, you'll add two layers to the map in ArcGIS Pro: a layer containing points that represent store locations and a layer containing points that represent customer locations. Next, you'll use the Assign Customers by Distance geoprocessing tool to identify which customers are likely to shop at which stores.

Creating an association between each customer and their closest store allows you to define markets using actual data rather than best guesses. In tutorial 17-2, you'll define these markets using the Generate Customer Derived Trade Areas geoprocessing tool.

## Open the project and prepare data points

1. In your browser, navigate to link.esri.com/Top20BA/Data.
2. Download the **Chapter 17 - Distance analysis** item. Extract the zip folder to **C:\Top20BusinessAnalyst**.

3. In the **Ch 17 - Distance analysis** folder, double-click **Distance analysis.aprx** to open the project in ArcGIS Pro.

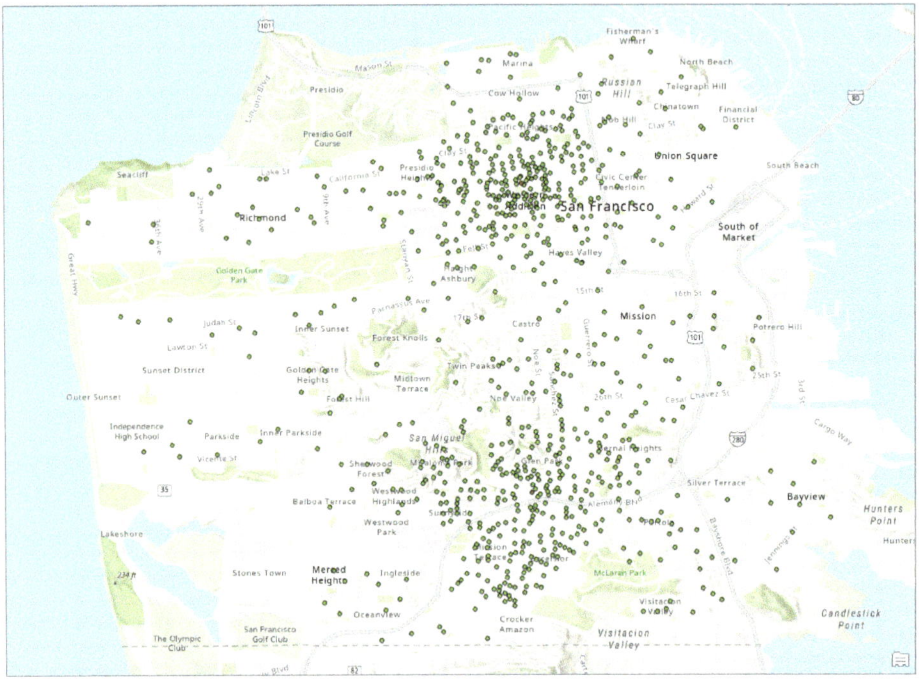


The map has two point layers that represent a sample dataset of stores and customers in San Francisco, California. To prepare for the analysis, you'll symbolize the points so that they are easier to distinguish from each other.

4. In the **Contents** pane, uncheck the box next to the **sf_customers_layer** to turn the layer off.

   The map has only two points representing the stores.

5. Beneath **sf_stores_layer**, click the circle symbol.

   The Symbology pane appears, showing symbol options for point locations. You'll change the symbol for the two store points so that they're easier to distinguish from the customer points.

6. In the **Symbology** pane, in the search field of the **Gallery** tab, type Esri pin and press Enter.

7. In the **ArcGIS 2D** section, click **Esri Pin 1**.

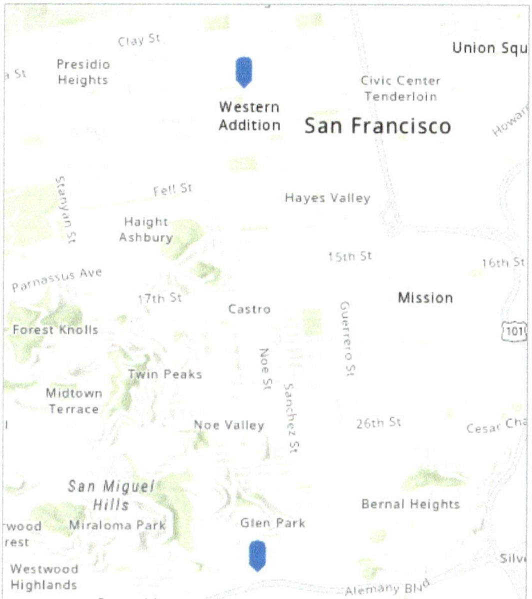


The symbols for the two store points are changed to the new symbol.

8. In the **Contents** pane, turn on **sf_customers_layer**.

   Next, you'll assign each customer to one of the two store locations.

## Assign customers to stores

The Assign Customers By Distance tool allows you to create an association between each customer and their closest store, based on a selected distance type. For the two San Francisco stores, you'll create an association based on driving distance.

9. On the ribbon, click the **Analysis** tab. In the **Geoprocessing** group, click **Tools**.

10. In the **Geoprocessing** pane, search for and open the **Assign Customers By Distance** tool.

11. In the **Assign Customers By Distance** tool, apply the following settings:

    - For **Input Customer Features**, select **sf_customers_layer**.
    - For **Input Store Features**, select **sf_stores_layer**.
    - For **Store ID Field**, select **NAME**.
    - For **Output Feature Class** and **Field Name**, leave the defaults.
    - For **Distance Type**, select **Driving Distance**.
    - For **Distance Units**, leave the default **Miles**.

12. Click **Run**.

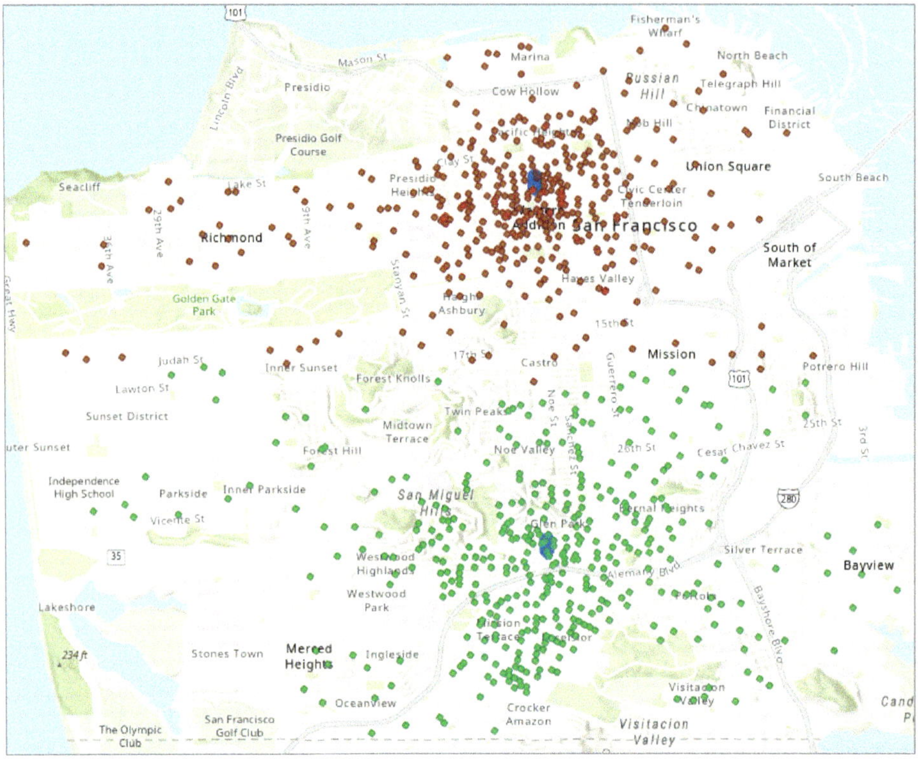


The points are now color-coded to show which store they are associated with based on driving distance. There is a new layer in the Contents pane that contains this data and displays a legend, denoting points as assigned to store 1 or store 2. Points in red are assigned to store 1 (the store at the top of the map), and points in green are assigned to store 2 (the store at the bottom of the map).

Now that you have assigned each customer point to a store based on their proximity when driving, you can create trade areas on the map for each store.

# Tutorial 17-2: Create trade areas based on customer distribution

In this tutorial, you will create trade areas as polygons on the map, showing the market for each store based on the percentage of customers. You will then create a second set of trade areas that show the same market but based on customer sales.

Customer-derived trade areas capture a specified percentage of the associated customers for each location. In the second example, you'll see trade areas based on the amount of sales from the store's customers. In this way, you will be able to see the area from which 80% of each store's sales is drawn.

## Create trade areas based on customers

You will create trade areas that depict 80% of each store's customers.

1. In the **Geoprocessing** pane, click the back button to return to the search. Search for Generate Customer Derived Trade Areas and open the tool.
2. In the **Generate Customer Derived Trade Areas** tool, apply the following settings:
   - For **Stores**, select **sf_stores_layer**.
   - For **Store ID Field**, select **STORE_ID**.
   - For **Customers**, select **sf_customers_layer_AssignCustomersByDistance**.
   - For **Associated Store ID Field**, select **STORE_ID**.
   - For **Output Feature Class**, type sf_stores_layer_80pct_of_customers.
   - For **Method**, select **Detailed**.
   - For **Radii (%)**, type 80.
   - Leave all other settings as is.

3. Click **Run**.

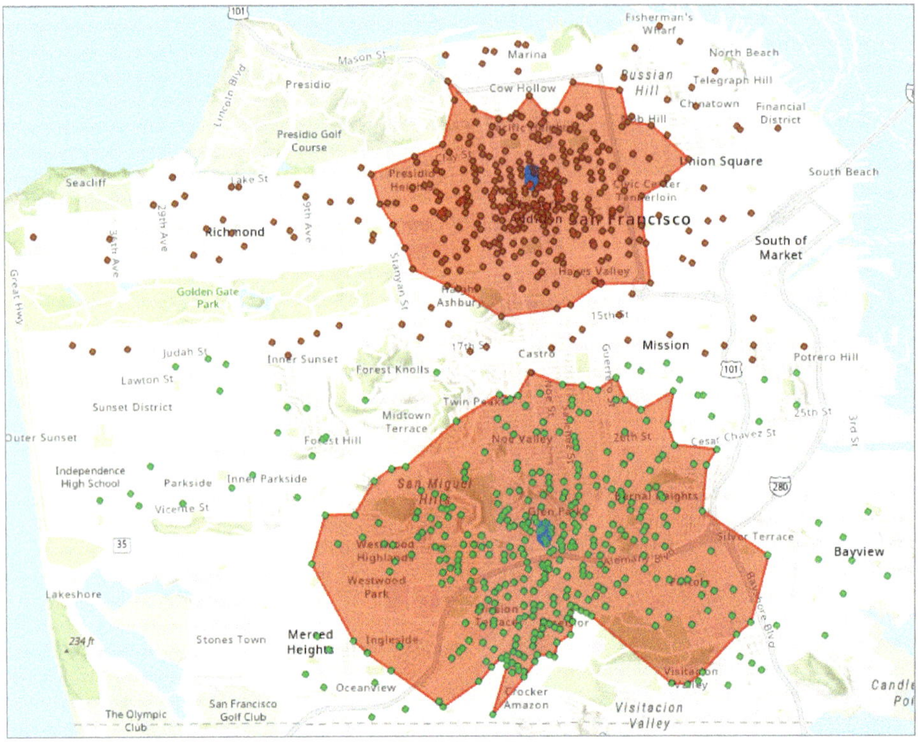

The trade areas are created, showing a polygon containing 80% of each store's customers.

## Create trade areas based on sales

You will create trade areas that depict 80% of each store's sales.

4. In the **Generate Customer Derived Trade Areas** tool, apply the following settings:

   - For **Output Feature Class**, type sf_stores_layer_80pct_of_sales.
   - For **Customer Aggregation Type**, select **Weight**.
   - For **Customer Weight Field**, select **SALES**.
   - Leave all other settings the same.

5. Click **Run**.

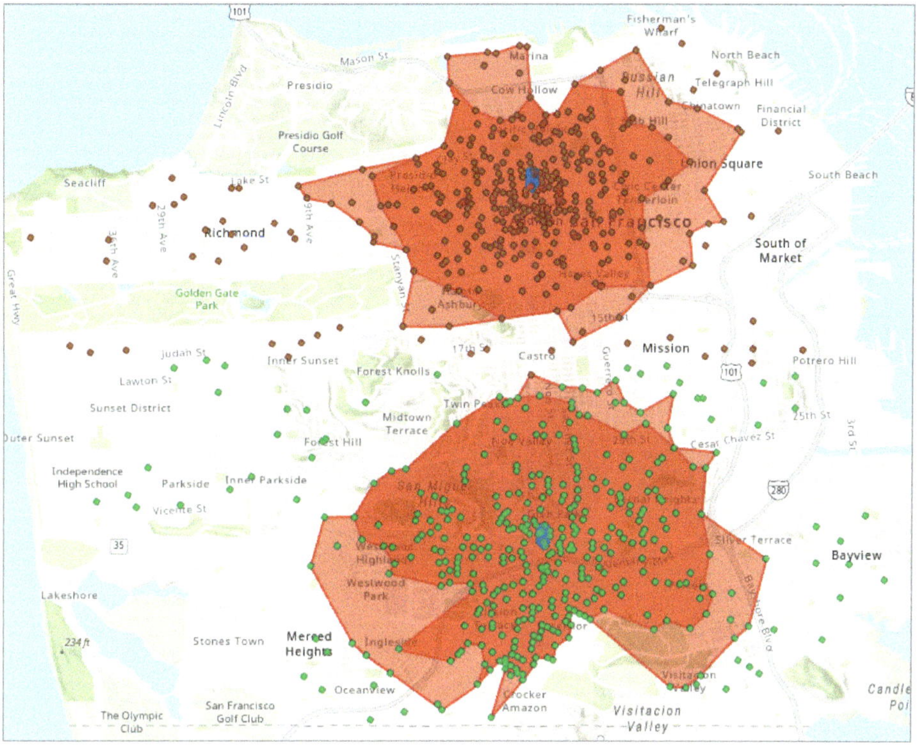


The trade areas are created, showing a polygon containing the customers that generate 80% of each store's sales.

The second set of trade areas is overlaid on top of the first set, using the same color. This makes the map slightly difficult to interpret, so you'll change the color of the sales polygons.

6. In the **Contents** pane, under the **sf_stores_80pct_of_sales** layer, click the rectangle symbol.

   The Symbology pane appears.

7. In the **Symbology** pane, click the **Properties** tab. Under **Appearance**, apply the following settings:

   - For **Color**, select **Apatite Blue**.
   - Click **Color** > **Color properties** and change the **Transparency** to 50%.
   - For **Outline color**, select the same color you chose for the polygons but do not adjust the transparency.

8. At the bottom of the **Symbology** pane, click **Apply**.

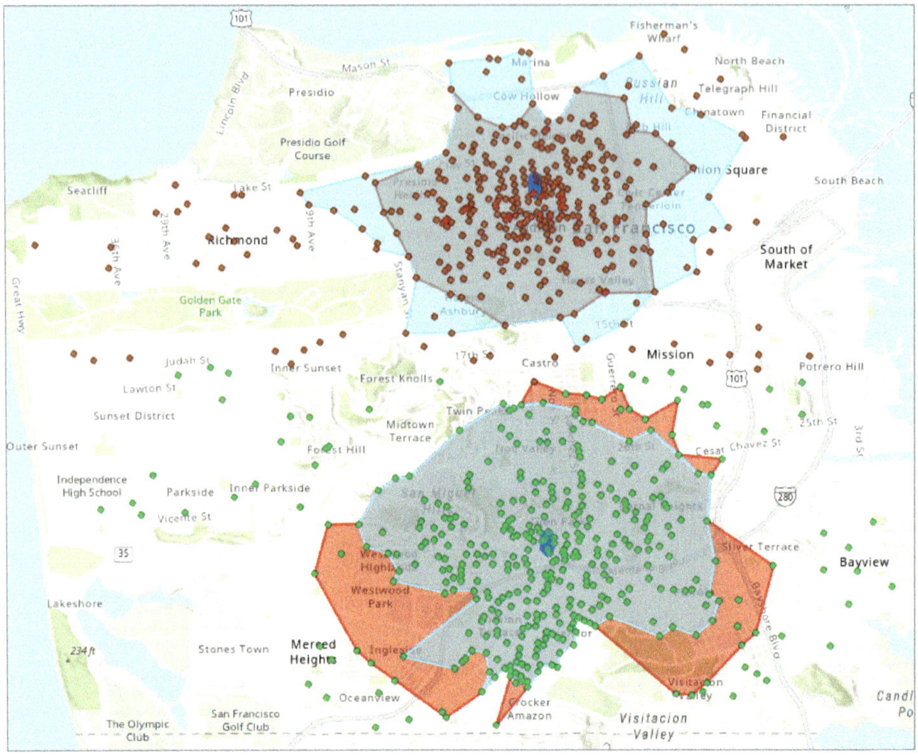


The sales polygons now reflect the updated symbology, allowing you to more easily see the relationship between the sets of trade areas you created. The trade areas based on sales are in blue, and the trade areas based on customers are in red.

The map now clearly shows that for store 1, the trade area based on sales is somewhat larger than the trade area based on customer count. For store 2, the trade area based on sales is smaller than the trade area based on customer count. What can we interpret from these differing trade areas?

This map may help each business prioritize their outreach and messaging based on different types of customers. Store 2 could target "loyal customer" communications on the sales-based trade area, focusing on the customers who make up the bulk of their revenue. Store 1 might cast a wider net in their communications or decide to reconfigure their business plans to cultivate more earnings from a smaller number of customers nearby.

## Summary

In this chapter, you mapped customer data and store locations. You then used these layers to create trade areas based on customer counts and sales totals.

### Workflow

1. Prepare data points for distance analysis.
2. Assign each customer to one of the two stores.
3. Create trade areas based on customer count and sales data.
4. Symbolize the map to better understand the trade areas.

# CHAPTER 18

# Creating advanced drive times

## Objectives

- Examine the distribution of the customer base and how spending at the mall varies across a trade area.
- Generate and assess the differences in trade area shape and customer composition.
- Visualize and examine the value of trade area techniques to best represent the underlying customer distribution and patterns.

## Introduction

Business Analyst Pro includes many trade area creation tools, which perform different analyses based on the type of data and desired trade areas. These include customer-based trade areas, as discussed in chapter 1, distance- and time-based trade areas (multiple chapters), and threshold trade areas (chapter 12). Each of these techniques uses a defined set of inputs, including the following:

- An individual facility, such as a store or set of locations
- A set of customers, who may visit a location or are connected to specific locations in some way
- A desired amount of an attribute, such as retail spending, where the trade area grows until the threshold is reached

- A combination of these inputs
- Other location or customer metrics, such as spending or sales

This chapter combines many of the trade area techniques discussed above and in previous chapters to describe how an organization can better model its customers, make more realistic and intuitive trade areas that mirror drive-time techniques, and learn how travel times and distances are impacted by traffic pattern change.

# Tutorial 18-1: Create trade areas based on customers and spending

To provide a realistic scenario, you will study customers that are captured in a store's Customer Relationship Management (CRM) system, which includes location and the total amount spent over the past year. The business is based in a retail mall in San Bernardino in Southern California. Using this data, you will create a range of trade areas that support different business agendas and needs for the store, including identifying where you might find more customers for planning, marketing, and outreach strategies to increase sales per customer visit.

In this tutorial, you'll use a predefined trade area created with the Generate Customer Derived Trade Areas geoprocessing tool using 50% of the customer base. This will be compared with two new trade area outlines that are based on 80% of the total spend and 80% of the customer base. You will compare the differences between these shapes and calculate the differences in spending between them to understand the value of different inputs to the customer-derived trade area process.

## Download and open the project

1. In your browser, navigate to link.esri.com/Top20BA/Data. Download the **Chapter 18 - Advanced trade areas** item.

2. Extract the **Ch 18 - Advanced trade areas** folder to **C:\Top20BusinessAnalyst**.

> **Tip:** This book uses **C:\Top20BusinessAnalyst** for the folder location, but you can use any folder.

3. In the **Ch 18 - Advanced trade areas** folder, double-click **Advanced Trade Areas.aprx** to open the project in ArcGIS Pro.

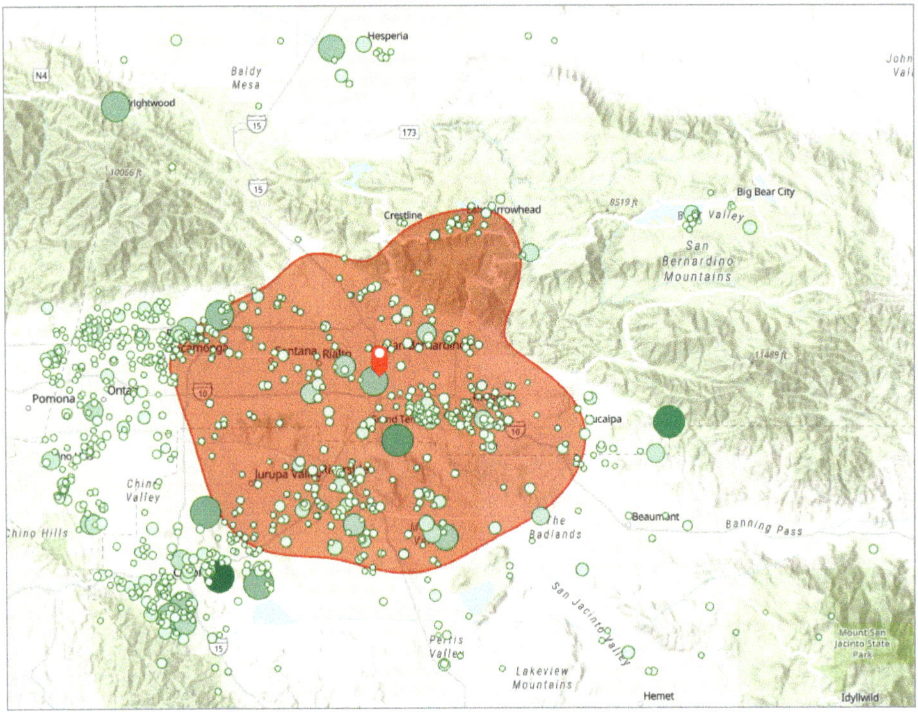


Once your project has loaded, you will see a map that shows the Inland Center Mall customers symbolized by individual customer spending and a smooth trade area created using Generate Customer Derived Trade Areas for 50% of the total number of customers.

The store location is represented by a red pin and customer spending is represented by green circle polygons. Higher spending at the mall is shown as larger and darker-colored circles. Visually, you can identify distinct clusters. Many of the highest spending customers are outside the trade area, which was created based on the number of customers rather than how much the customers spent.

## Create two new trade areas using 80% of the customer base and aggregate spending

The store wants to delineate the line for 80% of their customer base, assuming that they are the best customers, but they are unsure of whether to use the number of customers or the spending amount. Leadership believes that there are higher-spending customers very close to the 50% cutoff, which staff say is the maximum distance customers will travel. As an analyst, you will provide more insight to the leadership team to back up assumptions or provide additional facts and context that can lead to more informed decision-making and action.

To start your analysis, you will create a new set of customer-derived trade areas, or CDTAs, using 80% of the total number of customers and 80% of customer spending. This will be used to generate new statistical insights and shape subsequent analysis.

4. On the ribbon, click the **Analysis** tab. In the **Geoprocessing** group, click **Tools**.
5. In the **Geoprocessing** pane, search for and open the **Generate Customer Derived Trade Areas** tool.

   This CDTA will be based on the total number of customers.
6. In the **Generate Customer Derived Trade Areas** tool, apply the following settings:
   - For **Stores**, select **Inland_Mall**.
   - For **Store ID Field**, select **STORE_ID**.
   - For **Customers**, select **Inland_Mall_Customers**.
   - For **Associated Store ID Field**, select **STORE_ID**.
   - For **Output Feature Class**, type Inland_Mall_Customers_CDTA_Count.
   - For **Method**, select **Detailed with smoothing**.
   - For **Radii (%)**, type 80.
   - For **Customer Aggregation Type**, select **Count**.
7. Click **Run**.

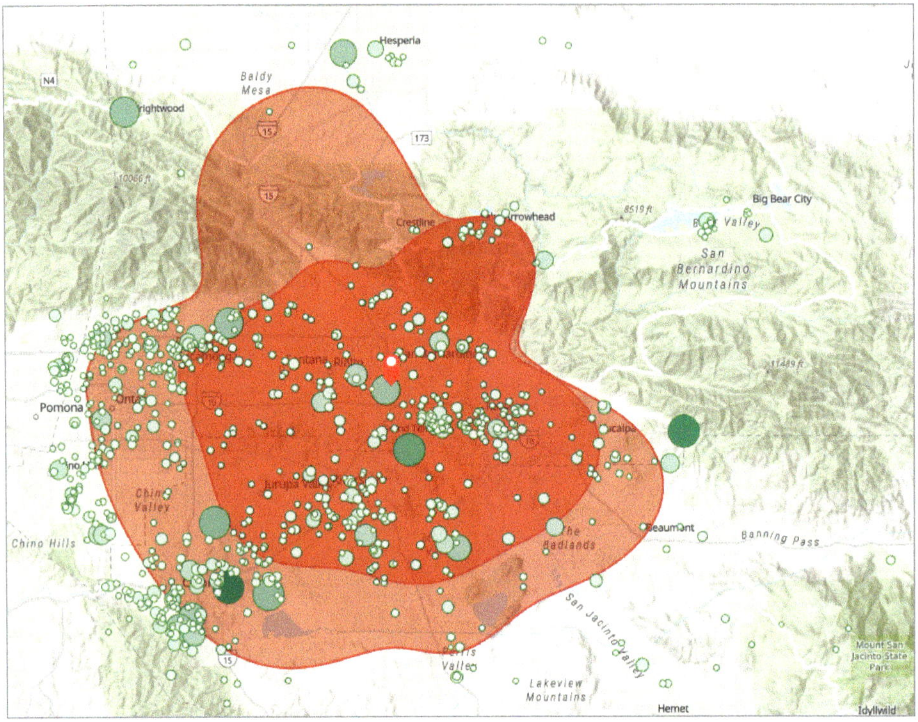


You now have a new CDTA that is larger than the original 50% of customers trade area. It includes additional customers and some of the highest spending clusters, especially in the southwest and northwest of the trade area.

You will repeat the calculation of a CDTA based on the total spending in dollars.

8. In the **Generate Customer Derived Trade Areas** tool, apply the following settings:

   - For **Stores**, select **Inland_Mall**.
   - For **Store ID Field**, select **STORE_ID**.
   - For **Customers**, select **Inland_Mall_Customers**.
   - For **Associated Store ID Field**, select **STORE_ID**.
   - For **Output Feature Class**, type Inland_Mall_Customers_CDTA_Weight.
   - For **Method**, select **Detailed with smoothing**.
   - For **Radii (%)**, type 80.
   - For **Customer Aggregation Type**, select **Weight**.
   - For **Customer Weight Field**, select **DOLLARS**.

9. Click **Run**.

10. In the **Contents** pane, turn off the **Inland_Mall_Customers_CDTA_Count layer**.

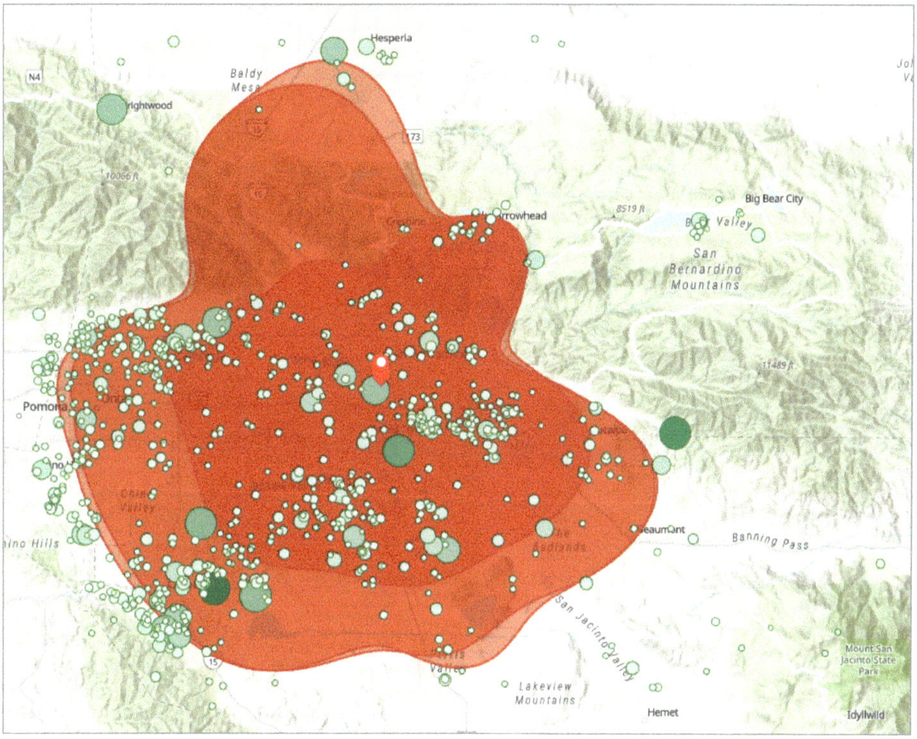


The CDTA based on total amount of spending is created, which extends beyond the edges of the 80% CDTA based on total count of customers, suggesting that there are higher-spending customers who live farther away from the store and that high amounts of spending are not found close to the store.

Note that the new CDTA extends farther north but neither of the 80% CDTAs captures all the highest spending customers, which are the largest, darkest green circles. One conclusion the store management might make is that these customers that come from far away may have different behaviors. For example, they may spend more in a single trip, which is for a special purpose, traveling farther than the local customers they consider to be core to the business.

11. On the map, click the **Inland_Mall_Customers_CDTA_Weight** layer. View the pop-up. Scroll down to find the **Count** and **Weight** attributes.

For the 80% CDTA that uses dollars spent as the weight, the pop-up summarizes the number of points in the CDTA as 664 and the total dollars spent as

$690,849.52. This means that, on average, each customer spends $1,040 at this particular Inland Center store.

## Exclude the outlying customers in the trade area calculation

You will generate another CDTA using dollars as the weight but exclude customers that are outliers based on a distance of 25 miles, which is the approximate radius of the amoebic-looking CDTA.

12. In the **Generate Customer Derived Trade Areas** tool, apply the following settings:
    - For **Stores**, select **Inland_Mall.**
    - For **Store ID Field**, select **STORE_ID.**
    - For **Customers**, select **Inland_Mall_Customers.**
    - For **Associated Store ID Field**, select **STORE_ID.**
    - For **Output Feature Class**, type Inland_Mall_Customers_CDTA_Weight_No_Outliers.
    - For **Method**, select **Detailed with smoothing**.
    - For **Radii (%)**, type 80.
    - For **Customer Aggregation Type**, select **Weight**.
    - For **Customer Weight Field**, select **DOLLARS**.
    - Check the box for **Exclude Outlying Customers**.
    - For **Customer Weight Field**, select **DOLLARS**.
    - For **Cutoff Distance,** type 25.
    - For **Distance**, select **US Survey Miles**.
13. Click **Run**.
14. In the **Contents** pane, turn off the **Inland_Mall_Customers_50Percent_DetailedwSmoothing** and **Inland_Mall_Customers_CDTA_Count** layers.

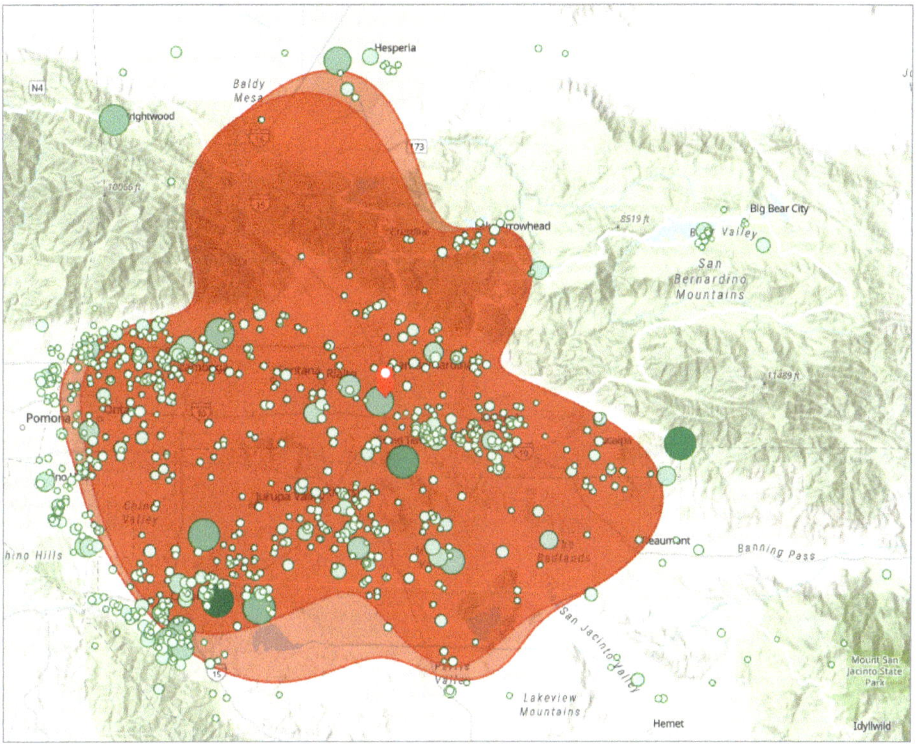


By removing the most distant customers that are more than 25 miles away, the new CDTA is more compact but is still quite similar to the 80% weighted trade area using dollars spent and no cutoff. This indicates that for this customer base, travel distance and spending are not highly correlated and that the difference in spending patterns is more likely due to the demographics of each customer and how well the store is merchandized for the customer base.

## Tutorial 18-2: Create an approximate drive-time trade area

The CDTA created using the detailed with smoothing method provides an organic outline for the CDTA, but for many analysts, managers, and business leaders, its rounded outline does not align with their perception of what a store-based trade area should be. Most analysts are used to network-based drive-time or drive-distance trade areas with a more typical jagged shape stretched along highways and branching into neighborhoods. The Generate Approximate Drive Times geoprocessing tool can convert the organic CDTA into one that is the best approximation of a drive time starting from the store. As a user, you can control the degree to which the new approximate drive-time trade area (ADTTA) matched the original CDTA by controlling the threshold of difference. By default, this is 5%, which balances performance with accuracy.

In this section, you are going to create the new ADTTA and compare how many customers are within it using the Summarize Within geoprocessing tool.

1. In the **Contents** pane, adjust the visibility of the layers so that only the following layers are turned on:

    - **Inland_Mall**
    - **Inland_Mall_Customers**
    - **Inland_Mall_Customers_50Percent_DetailedwSmoothing**
    - **Basemaps**

2. In the **Geoprocessing** pane, click the back button to return to the search.

3. Search for and open the **Generate Approximate Drive Times** tool.

4. In the **Generate Approximate Drive Times** tool, apply the following settings:

    - For **Input Features**, select **Inland_Mall_Customers_50Percent_DetailedwSmoothing**.
    - For **Output Feature Class**, type Inland_Mall_Customers_50Percent_ADTTA.
    - For **Distance Type**, make sure it is set to **Driving Time**.
    - For **Distance Units**, make sure it is set to **Minutes**.
    - For **Store Layer**, select **Inland_Mall**.
    - For **Store ID Field**, select **STORE_ID**.
    - For **Associated Store ID Field**, select **STORE_ID**.

5. Click **Run**.

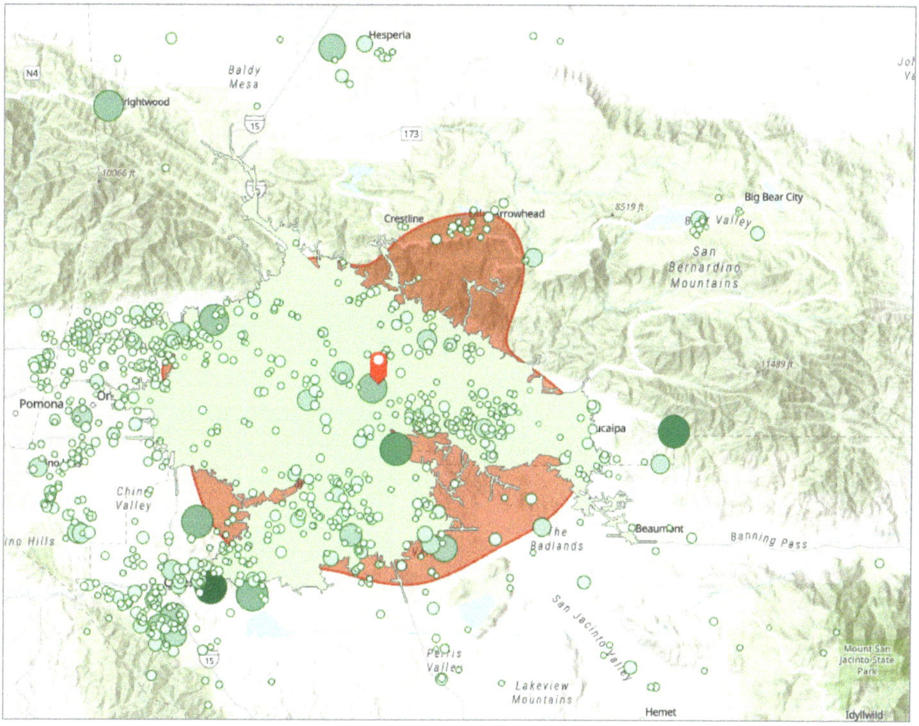


You will now have a new ADTTA on the map with the network trade area centered on the Inland Center Mall.

## Summarize the customers within the new ADTTA

The new ADTTA looks different from the amoebic, smooth CDTA, but how well does it perform in terms of capturing total customers? One quick way to test it is to use the Summarize Within geoprocessing tool to count the number of customers in the ADTTA.

6. Return to the search in the **Geoprocessing** pane. Search for and open the **Summarize Within (Analysis Tools)** tool.
7. In the **Summarize Within** tool, apply the following settings:
   - For **Input Polygons**, select **Inland_Mall_Customers_50Percent_ADTTA**.
   - For **Input Summary Features**, select Inland_Mall_Customers.
8. Click **Run**.

9. On the map, click on the summary layer to bring up the pop-up.

   In the pop-up, you will see that the average driving time was calculated from the store location to the most common edge of the CDTA, which is 25.88 minutes. You will also see that the ADTTA has 386 points compared with 401 in the original CDTA.

10. Close the pop-up.

> **Tip:** When creating a CDTA and a child ADTTA, the number of points in the original layer will be in the **Count** attribute in the pop-up. If the CDTA was created using a weight such as spending in dollars, the **Count** attribute will reflect the total number of captured points, typically customers, and the **Weight** field will reflect the total aggregate weight, such as the total spending for all points in the CDTA.

When comparing the CDTA and ADTTA trade areas, the ADTTA has 15 points missing compared with the CDTA. This is because the original CDTA was created using a geometric technique, which does not consider geographic distribution or the underlying road network.

When you're using the road network, the best approximation of the drive time from the store is 25.88 minutes. The ADTTA and CDTA generally align well, validating the quality of the underlying algorithm used in the Generate Approximate Drive Times tool. However, some locations are beyond the 26-minute range, notably the mountain community around Lake Arrowhead to the northeast and Lake Perris neighborhoods to the southeast.

## Create a new CDTA using the drive-time option

The Generate Customer Derived Trade Areas tool has an advanced option that uses drive times and drive distances to more accurately reflect the underlying road network, barriers, and topographic features, rather than using geometric shapes.

11. In the **Geoprocessing** pane, search for and open the **Generate Customer Derived Trade Areas** tool.
12. In the **Generate Customer Derived Trade Areas** tool, apply the following settings:
    - For **Stores**, select **Inland_Mall**.
    - For **Store ID Field**, select **STORE_ID**.

- For **Customers**, select **Inland_Mall_Customers**.
- For **Associated Store ID Field**, select **STORE_ID**.
- For **Output Feature Class**, type Inland_Mall_Customers_CDTA_DriveTimes.
- For **Method**, select **Threshold Drive Times**.
- For **Radii (%)**, type 50.
- For **Customer Aggregation Type**, select **Count**.
- Make sure the **Exclude Outlying Customers** box is unchecked.

13. Click **Run**.

14. In the **Contents** pane, turn off all layers except **Inland_Mall**, **Inland_Mall_CDTA_DriveTimes**, and the basemaps.

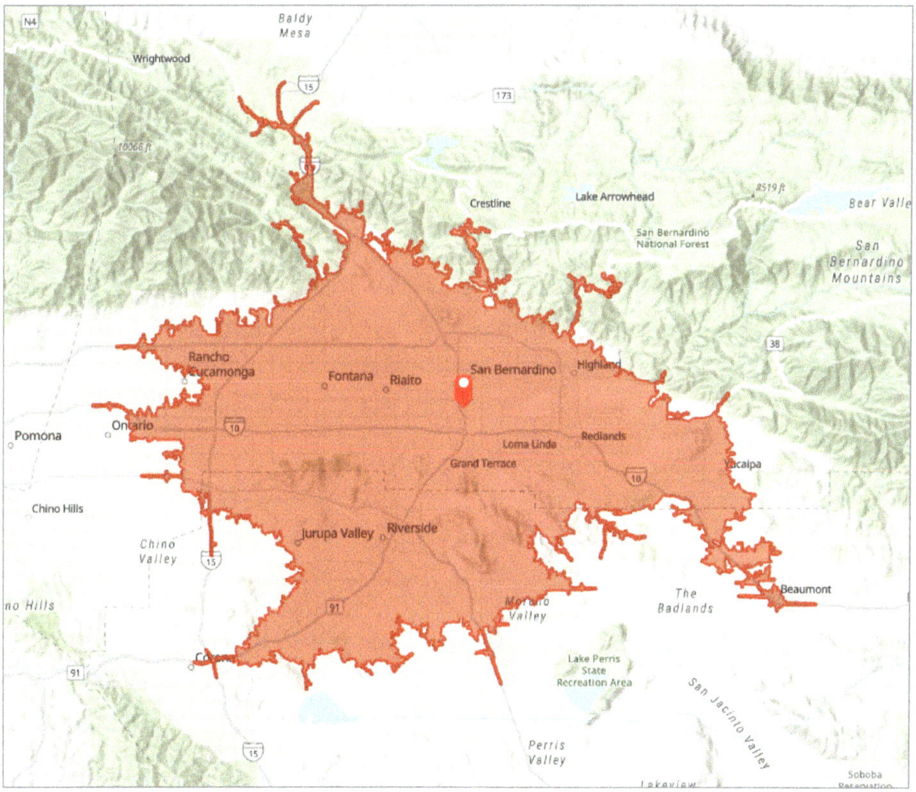


15. On the map, click the new **Inland_Mall_Customers_CDTA_DriveTimes** feature layer to view the pop-up.

    The 50% threshold drive time (the Radius attribute) is 26.28 minutes, only 40 seconds more than the approximation generated for the ADTTA trade area.

## Additional use cases for ADTTAs

### Privacy and hiding customer locations

Many organizations, such as banks and medical facilities, have customers or clients with specific privacy requirements that their locations cannot be shared or shown on a map. At the same time, the organization may have a need to create realistic customer-based trade areas for other teams or for analysis.

Creating an ADTTA does not require the underlying points (customers, clients, patients, and so on), but just the location of the facility or the center of the CDTA. This helps protect data privacy and personal information while also creating accurate models of customer distributions that can be used to create metrics, such as the number of customers served, percent coverage based on time travel, or the longest or farthest distance a visitor to the facility will have to travel.

### Summarizing places, such as territories, counties, and zip codes

Many GIS users and the public find it hard to understand measures such as square miles when discussing places. The ADTTA tool in the Trade Areas toolbox can be used to create a more understandable and communicable metric for geographic polygons, such as counties, states, zip codes, or towns and cities.

In the following example, an ADTTA has been created for Orange County, California. The result is a 31-minute drive time, from the center of the county, showing how the mountains, bays, and beaches might impact travel across the region.

The ADTTA also provides valuable insight to the proximity of communities in Los Angeles County to Orange County and how the State Route 91 (SR 91) freeway makes Corona, in Riverside County, as accessible as Laguna Niguel.

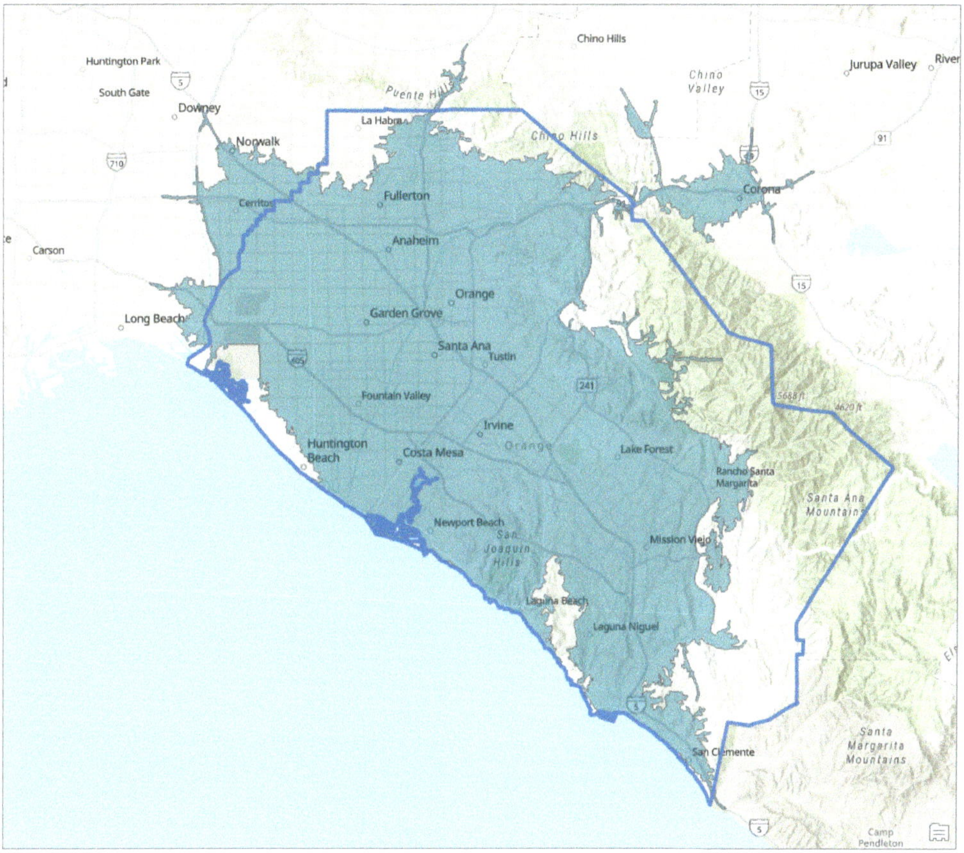


## Summary

In this chapter, you used Business Analyst Pro to model customer interactions with a shopping center and create trade areas using various techniques to include a proportion of the customer base and the total aggregated spending by customers, as well as geometric- and geographic-based analysis models.

### Workflow

1. Load and open an example project.
2. Create trade areas around the store using different techniques.
3. Perform comparison analysis between the different tools and learn how to apply these techniques to different use cases.
4. Present the results on a map and in pop-ups.

# CHAPTER 19

# Using target marketing to target key customer segments

## Objectives

- Use the ArcGIS Tapestry segmentation system to analyze and understand customers in a specific market.
- Quantify the relative and absolute distribution of each Tapestry segment to better understand the market and customer characteristics.
- Use a four-quadrant analysis to classify segments into distinct target groups for which marketing strategies can be developed.
- Generate market penetration maps and assess which neighborhoods are most likely to respond to marketing efforts.

## Introduction

Target marketing is the process of analyzing customers within a market to generate key insights that will be used to create better marketing campaigns. The workflow uses geographic distributions of customers compared with their neighbors and the market area to guide the user in their work. It is an essential tool to find your best customers and those that you can develop deeper relationships with to help drive engagement and future sales.

The Target Marketing Wizard in Business Analyst Pro provides:

- Step-by-step guidance for users to profile, visualize, and analyze customers
- Grouping tools that assign customers into four distinct target groups
- Identification of areas with high potential engagement with the campaign
- Recommendations for media strategies to reach those customers

The wizard makes maps and reports that are ready for the boardroom. It also helps users make marketing plans that maximize return on investment.

In this chapter, you are a marketing expert for a large car dealership in Las Vegas, Nevada. You have a mailing list of potential, current, and past customers that has been converted into customer points through the geocoding of addresses. You have subsequently enriched the layer of customer points with the dominant segment.

## Tutorial 19-1: Use the Target Marketing Wizard

Using target marketing, you will explore the distribution of customers and compare them with the Las Vegas market. This will help you identify where you should concentrate your marketing outreach strategies that aim to increase sales at the dealership.

The Target Marketing Wizard uses five steps to turn raw customer points into informed decisions:

1. Create a Customer Profile based on the customers' Tapestry segments to reveal which segments engage with a product or service the most.
2. Create a Market Area Profile for a specific area. You can compare your customer segments to see which areas have more potential for growth and which areas are not performing well.
3. Perform a Four Quadrant Analysis to categorize customers into Core, Developmental, Niche, and Monitor target groups.
4. Map the target groups to reveal where to target marketing investments based on concentrations of Core and Developmental target groups.
5. Create reports that summarize customer groups, market potential, and media habits. This will help you make better marketing decisions.

> *Important: Target Marketing tools are supported only with locally stored ArcGIS Business Analyst datasets. If you do not have a local dataset, skip this tutorial and go to tutorial 19-2.*

## Download and open the example project

1. In your browser, navigate to link.esri.com/Top20BA/Data. Download the **Chapter 19 - Target marketing** item.
2. Extract the **Ch 19 - Target marketing** folder to **C:\Top20BusinessAnalyst**.

   **Tip:** This book uses **C:\Top20BusinessAnalyst** for the folder location, but you can use any folder.

3. In the **Ch 19 - Target marketing** folder, double-click **Las Vegas Target Marketing.aprx** to open the project in ArcGIS Pro.

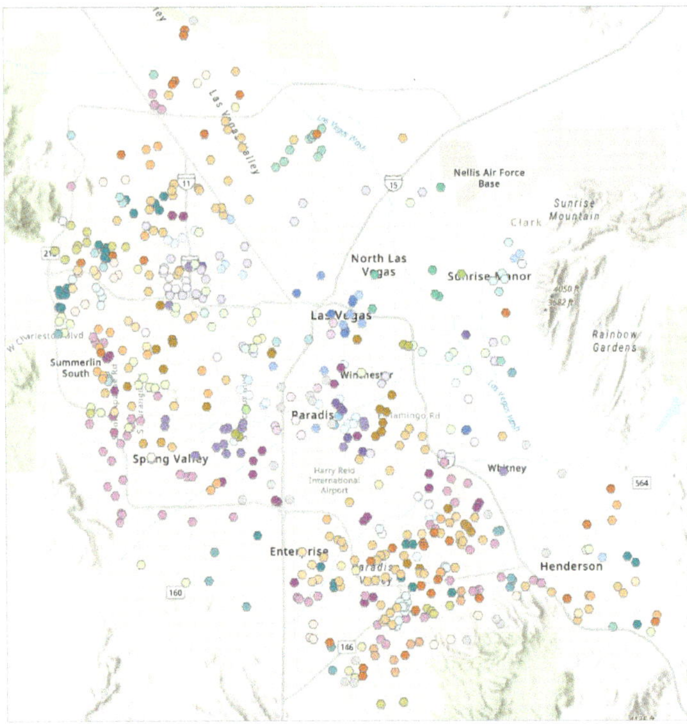


Once your project has been loaded, you will see a map that shows the distribution of customers symbolized by their individual Tapestry segments.

## An introduction to Tapestry

Tapestry is a geodemographic segmentation system that classifies US residential areas into distinct segments based on shared demographic, socioeconomic, and lifestyle characteristics. Neighborhoods with the most similar characteristics are grouped together, and neighborhoods showing divergent characteristics are separated.

For a broader view of the community, segments are summarized into LifeMode groups. Each LifeMode group has different groups that share basic demographic characteristics. These characteristics include age, income, type of household, and the number of people living in the area.

LifeMode groups are further divided into LifeStage groups, which reflect the age and family composition of the shared LifeMode groups. The LifeStage groups are called Contemporary Households, Family Portraits, and Mature and Retired Living. These groups reflect how adults change as they age, forming families while working and then retiring. They can be defined as follows:

- **Contemporary Households**—leaving home and establishing relationships,
- **Family Portraits**—forming families and engaging in career change and moving into different housing types, and
- **Mature and Retired Living**—older heads of households and individuals near or postretirement.

Overall, there are 60 segments in 12 LifeMode groups and three LifeStage groups. Each LifeMode is assigned a distinct brand color, making it easy to identify common segments within LifeMode groups. To view the chart in full, go to link.esri.com/Top20BA/LifeMode.

Each Tapestry segment is summarized based on predominant households and housing types, together with key demographics, such as household composition, diversity, age, and income level. Each segment has its own behavior patterns, such as where they shop or like to go. These patterns show the differences between each segment.

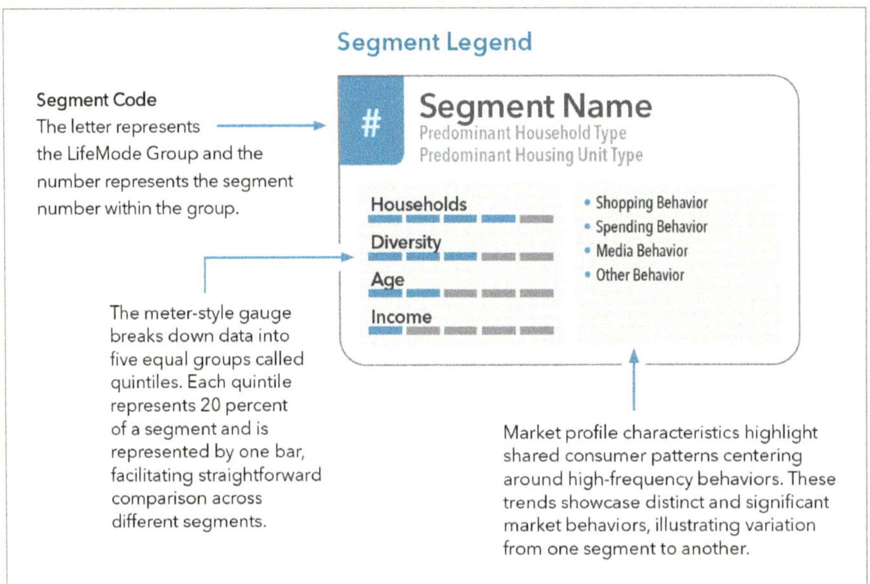


As an example, the Fresh Ambitions segment, which is in the Urban Threads LifeMode group, is made up of predominantly married couples or singles living alone. This segment is found in more dense urban areas, where they live in apartment blocks or smaller single-family homes.

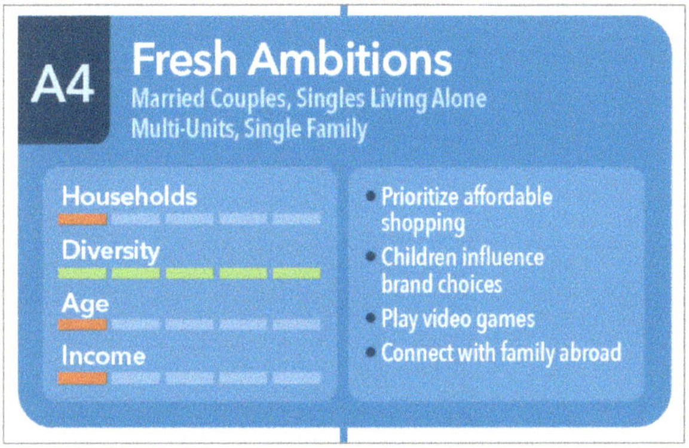


## Create a customer profile for the Las Vegas market

The aim of target marketing is to turn visualization into quantitative insight.

You will create a new customer profile from the customer feature layer. This includes 607 customers spread across the Las Vegas metro area. Using the map legend and symbology, you can see clusters of common customer segments and other areas where there are few customers.

4. On the ribbon, click the **Analysis** tab. In the **Workflows** group, click the **Business Analysis** gallery and click **Target Marketing Wizard**.

   The Target Marketing Wizard pane appears. The first step creates a customer profile with summary information.

5. For **Step One: Customer Profile**, apply the following settings:

   - For **Customer Layer**, select **Las_Vegas_Customers**.
   - For **Customer Segmentation Profile**, select **Create new profile** and type Las_Vegas_Customers_Target_Profile.
   - For **Segmentation Base**, make sure it is set to **Total Households**.
   - For **Volumetric Field**, make sure it is set to **<none>**.

6. Click **Next**.

   Using Tapestry segments, you now have a summary of customers. This will be used to compare the number of households in each segment across the Las Vegas market. It will give you a way to measure how different your customers are from the whole market.

## Create a segmentation profile for the Las Vegas market

The second step involves creating a segmentation profile for every block group in the Las Vegas market based on the number of households in each of the Tapestry segments. The segmentation profile is based on a market area—in this case, Las Vegas—which is a table of distributions of each segment.

This tutorial compares the types of customers and market segments in Las Vegas Metro to other areas. You could also use segmentation profiles based on other markets, such as Los Angeles, New York, the whole of Nevada, or even the United States. Each segmentation profile is the basis against which you understand and quantify your customers. It directs and assists your marketing strategy.

7. For **Step Two: Market Area Profile**, apply the following settings:

   - Make sure the **Market Area** is set to **Current Display Extent**.
   - For **Market Area**, make sure **Create new profile** is selected.
   - For **Create new profile**, type Las_Vegas_Base_Profile.

8. Click **Next**.

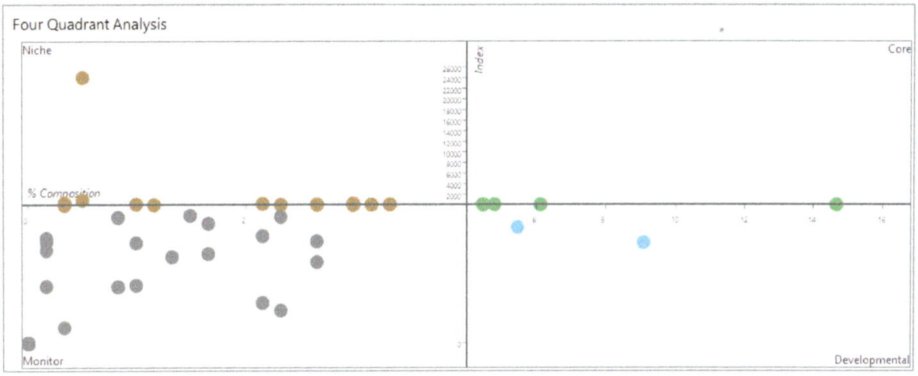


You now have customer and market profiles, which are combined to create the Four Quadrant Analysis chart.

## Four Quadrant Analysis chart explained

The Four Quadrant Analysis chart looks at the customer segmentation profile by comparing it to create indexes. The chart plots the segments that represent your customers based on percent composition (x-axis) and index (y-axis) values:

- **Percent Composition**—the percentage of your customers assigned to each segment based on their location.
- **Index**—the index measures the percentage of your customers assigned to a given segment against the percentage of people in that segment in the market area.

An index above 100 means that you are attracting customers from that segment at a higher rate than they occur in the market area. For example, if 10 percent of customers in the target profile and 5 percent of people in the base profile are assigned to segment x, the index is 200 (10 % / 5 % × 100). This indicates that you are attracting customers from that segment at twice the rate that they occur in the market area.

You can change the Index and Percent Composition values by typing them in the text boxes. You can also drag the x- or y-axis to change which segments fit into each of the four quadrants.

By default, Four Quadrant Analysis creates four target groups: Core, Developmental, Niche, and Monitor. Each group represents a different level of penetration and patronage:

- **Core target**—segments that contain a high percentage of customers and above-average indexes. These main customers engage with or spend more money on your brand than the general population in the area. *These segments represent loyalty and good opportunities to increase sales.*
- **Developmental target**—segments with a lot of customers, but that do not index above average. These segments are important because they represent a significant portion of your customers. However, they are consuming your product or service at a lower rate than the general population and especially when compared with the core target group. *These segments are the most likely to be converted into loyal customers.*
- **Niche target**—segments that represent a small percentage of customers but index above average, often at levels similar to or above those of the core target. These groups spend more money than the general population in their area, but they do so in small amounts. *Niche segments represent opportunities where a business can expand or explore new markets.*
- **Monitor target groups**—groups that don't make up a lot of your customers and have low average visits. These are consumer segments that do not purchase your product or service. *Usually, you won't use this group of segments in marketing campaigns. But you should still watch them, especially when they are close to moving from Monitor to Niche or Developmental target groups.*

9. In the **Four Quadrant Analysis** chart, click **Customer Profile** to view the customer segmentation profile.

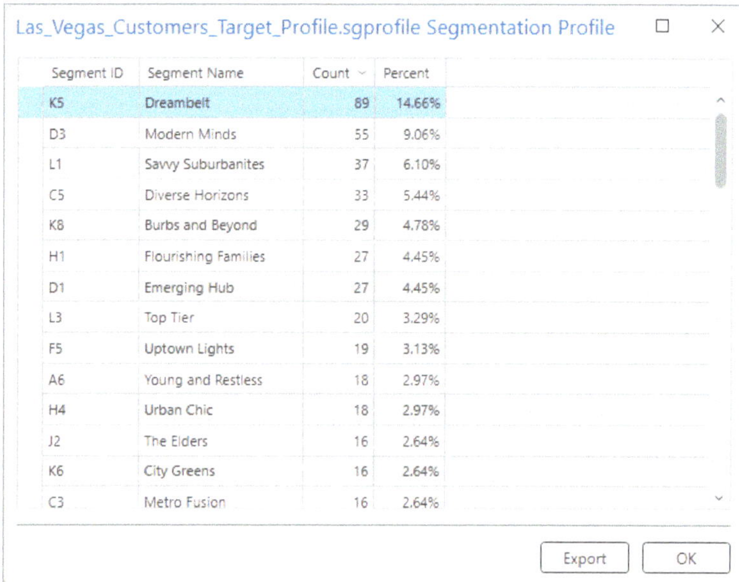


| Segment ID | Segment Name | Count | Percent |
|---|---|---|---|
| K5 | Dreambelt | 89 | 14.66% |
| D3 | Modern Minds | 55 | 9.06% |
| L1 | Savvy Suburbanites | 37 | 6.10% |
| C5 | Diverse Horizons | 33 | 5.44% |
| K8 | Burbs and Beyond | 29 | 4.78% |
| H1 | Flourishing Families | 27 | 4.45% |
| D1 | Emerging Hub | 27 | 4.45% |
| L3 | Top Tier | 20 | 3.29% |
| F5 | Uptown Lights | 19 | 3.13% |
| A6 | Young and Restless | 18 | 2.97% |
| H4 | Urban Chic | 18 | 2.97% |
| J2 | The Elders | 16 | 2.64% |
| K6 | City Greens | 16 | 2.64% |
| C3 | Metro Fusion | 16 | 2.64% |

The Las Vegas Customer Target Profile summarizes each of the segments as both counts and percentages. Of the 609 customers in Las Vegas, 89, or 14.66 percent, are in Tapestry segment K5 Dreambelt, and 55, or 9.06 percent, are in the D3 Modern Minds segment. This means that almost one-quarter of the customer base is concentrated in just two segments.

10. Click **OK** to close the profile.

## Explore the customer segmentation distributions in the Four Quadrant chart and the Target Groups pane

The third step is to look at and improve the Four Quadrant chart and see whether you should assign segments to different groups.

11. For **Step Three: Configure Target Groups**, under **Targets**, expand the **Core** and **Developmental** target groups to see the constituent segments in each of the target groups.

| Core (5) | Index | Percent Composition |
|---|---|---|
| D1: Emerging Hub | 225 | 4% |
| H1: Flourishing Families | 148 | 4% |
| K5: Dreambelt | 143 | 15% |
| K8: Burbs and Beyond | 203 | 5% |
| L1: Savvy Suburbanites | 163 | 6% |
| **Developmental (2)** | **Index** | **Percent Composition** |
| C5: Diverse Horizons | 93 | 5% |
| D3: Modern Minds | 80 | 9% |
| **Niche (15)** | **Index** | **Percent Composition** |
| **Monitor (39)** | **Index** | **Percent Composition** |

The Core target group includes five segments, which make up 34 percent of all customers. Developmental has two segments (Diverse Horizons and Modern Minds) that make up 14 percent of the customer base. These two groups are the most important for the marketing campaign and represent the most engaged customers. Combined, these seven segments compose almost half the customer base, at 48 percent.

12. Close the **Core** target group and expand the **Niche** target group.

You can see in the Four Quadrant chart and in the Niche target group, the Unclassified segment index is 24,570 because there are no customers in this group. Consequently, you'll want to exclude this segment by moving them to their own target group, which you'll call Exclude.

> **Tip:** When you do target marketing in your company, you usually want to exclude groups with no customers that also have an index of over 1,000.

13. Next to **Targets**, click the **Add** button to add a new target group.

14. Set the name of the new target group to Exclude.

15. From the **Niche** target group, drag the **Unclassified** segment to the new **Exclude** target group.

    With the new target group containing the outlier, you can now delete this group to rebalance the Four Quadrant chart.

16. Hover over the **Exclude** target group and click the red X to delete it.

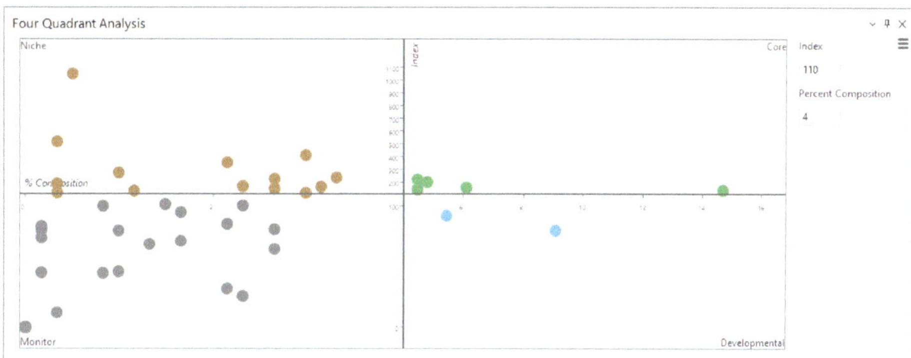


A new Four Quadrant Analysis chart is created without the Unclassified segment. The chart is much easier to read and understand because the index range, which represents the y-axis in the chart, now has a smaller range. This highlights that D5: Laptops and Lattes is another outlier with zero customers, but for the purposes of this tutorial, you can ignore this segment.

You can now see all five segments that make up the Core target group, and how three segments cluster near the intersection of the chart's axes. To see the names of each segment and its summary statistics, hover over the circle in the chart.

The Niche and Monitor target groups are similar, but they have fewer customers than the Core and Developmental target groups. This means you have classified the segments correctly. You can now move to the next step of analysis, which is mapping the segments to create a target, penetration, and target group layer. You can also map market potential and gaps.

17. Click **Next**.

**Tip:** If you accidentally close the Four Quadrant chart and want to view it again, use the **Menu** button in the upper right of the wizard pane for **Step 3: Configure Target Groups** to reopen it.

## Generate maps of the core targets and target group layers

18. For **Step Four: Mapping**, apply the following settings:

    - For **Geography Level**, make sure **US.BlockGroups** is selected
    - For **Mapping Layers**, expand **Target Layer** and make sure **CORE** is selected.
    - Uncheck the **Penetration Layer** box.
    - Make sure the **Target Group Layer** box is checked.
    - Uncheck the **Market Potential** box.
    - Uncheck the **Market Area and Gap Analysis** box.

19. Click **Next**.

## Create segmentation and developing marketing strategies reports

20. For **Step Five: Reporting**, under **Target Marketing Reports**, apply the following settings:

    - Check the **Segmentation Profile** box.
    - Check the **Developing Marketing Strategies** box.
    - Expand the **Developing Marketing Strategies list. Make sure the Core Target is set to Core. Make sure Developmental Target is set to Developmental.**
    - Make sure the **Market Potential** box is unchecked.
    - Make sure the **Market Area and Gap Analysis** box is unchecked.
    - In the **Export Format** area, make sure the **PDF** box is checked.

21. Click **Next**.

22. For **Step Six: Summary** pane, click **Finish**.

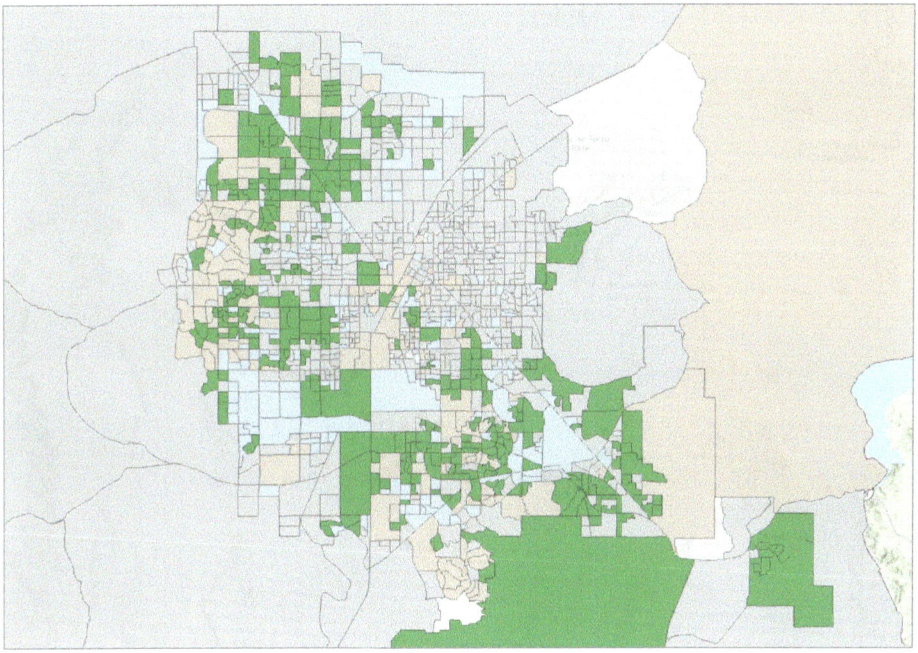

The wizard creates two additional feature layers in the Target Marketing group. There is a Target Layer (Core) that has block groups (which were chosen as the geography level in Step 4: Mapping) and a Target Group Layer. Each of the block groups has different colors for each target group.

## Tutorial 19-2: Gain insights from block group layers

In this tutorial, you will map the target layer using Tapestry symbology and filter the target group layers to explore the insights contained within the layers.

### For users who don't have a local dataset

If you do not have a local dataset, follow these steps to download the project:

1. In your browser, navigate to link.esri.com/Top20BA/Data. Download the **Chapter 19 - Target marketing (Segmented)** item.
2. Extract the **Ch 19 - Target marketing** folder to **C:\Top20BusinessAnalyst**.
3. In the **Ch 19 - Target marketing** folder, double-click **Las Vegas Target Marketing.aprx** to open the project in ArcGIS Pro.

## Symbolize block groups with Tapestry styling

1. In the **Contents** pane, uncheck **Target Group Layer** to turn off the layer.

   The target layer consists of two types of block groups, those that match your Core target group segments, shown in green and labeled "In" in the legend, and those that don't, shown in gray and labeled "Out."

   It is often easier to understand the pattern and distribution of target block groups by changing the symbology of the Out layer.

2. Under **Target Layer (Core)**, click the rectangle symbol for the **Out** sublayer.

   The Symbology pane appears.

3. In the **Symbology** pane, collapse the **ArcGISTapestry** section. Under **ArcGIS 2D**, click the first symbol, **Black Outline (1 pt)**.

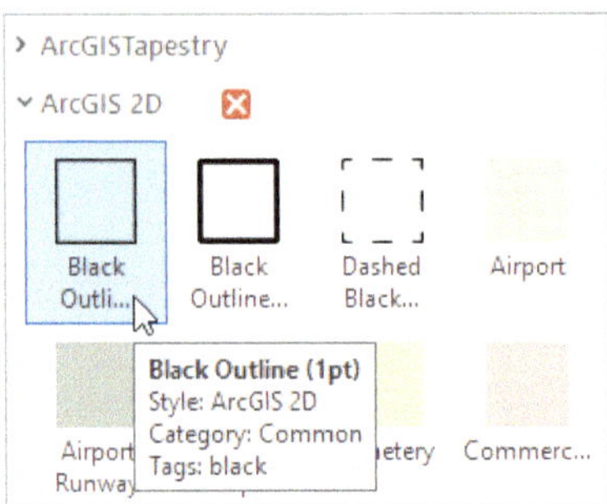


You now have a map that more helpfully displays the block groups that match the Core target group segments, which have customers, and those block groups that do not.

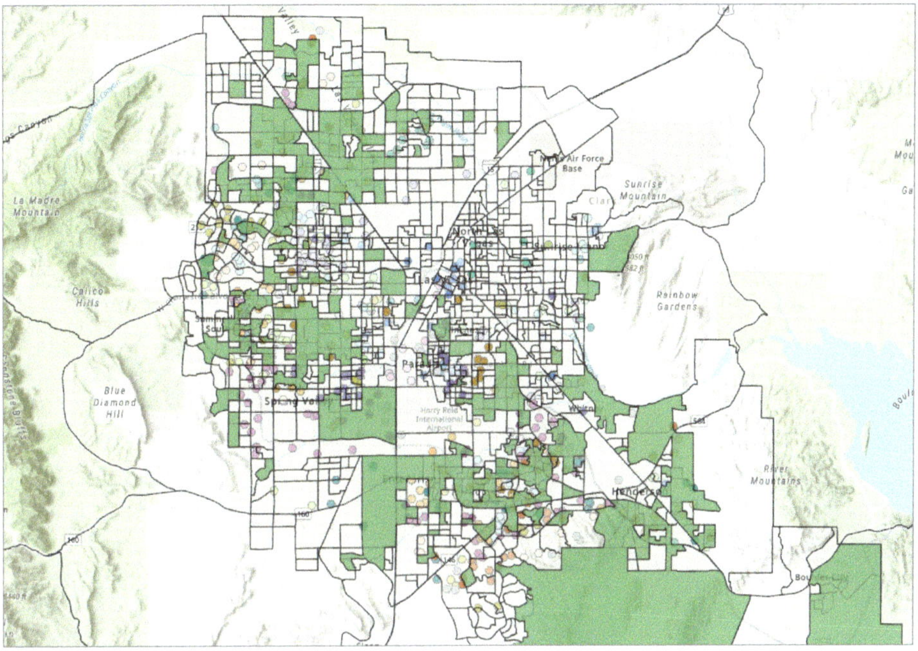


**Tip:** You can use the symbology **Properties** pane to further customize the outline color.

Although helpful, the map does not explain where each of the segments in your Core target group can be found. To do this, you will change the symbology of the In layer to match the brand colors used for each as shown in the underlying customer points. You'll use a style file specifically designed for Tapestry.

4. On the ribbon, click the **Analysis** tab. In the **Geoprocessing** group, click **Tools**.
5. In the **Geoprocessing** pane, search for and open the **Match Layer Symbology To A Style** tool.
6. In the **Match Layer Symbology To A Style** tool, apply the following settings:
   - For **Input Layer**, select **Target Layer (Core)**.
   - For **Match Values (Field or Expression)**, select **2025 Dom Tapestry Segment Name**.
   - For **Style**, select **ArcGISTapestry**.

7. Click **Run**.

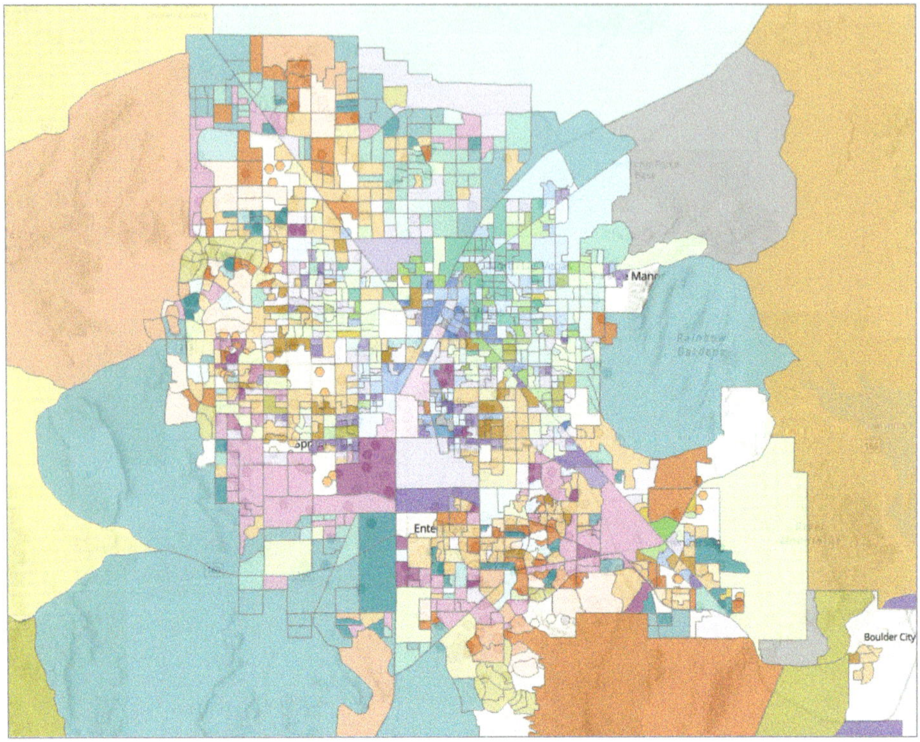


You now have a block group map symbolized with Tapestry colors.

## Filter block groups

Next, you need to filter out just those block groups that are in your target group.

8. In the **Contents** pane, right-click **Target Group (Core)** and click **Properties**.

9. In the **Layer Properties** window, click the **Definition Query** tab. lick **New definition query**.

10. For **Query 1**, for the **Where** field, select **Status**.

    The criteria operator automatically fills in as is equal to.

11. In the next empty cell, select **In**.

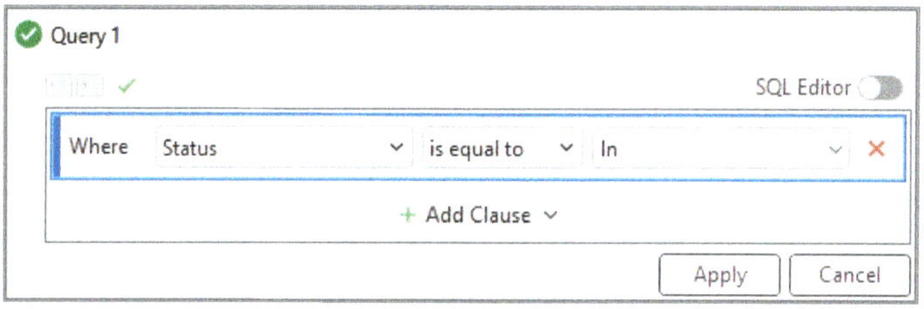


12. Click **Apply** and then click **OK**.

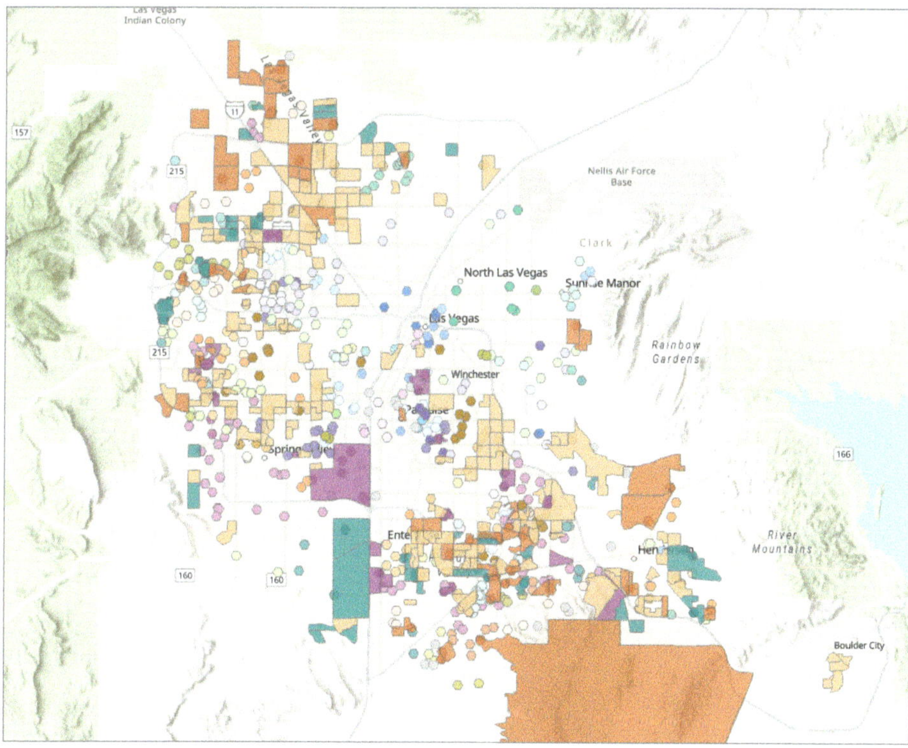


You now have a map with the Core target group, symbolized by the dominant Tapestry segment name. This allows you to navigate the map and see where you are missing target neighborhoods and what segments are missing despite them being in your customer database.

13. In the **Contents** pane, collapse the **Target Layer (Core)** legend. Turn off the layer.

14. Turn on **Target Group Layer**.

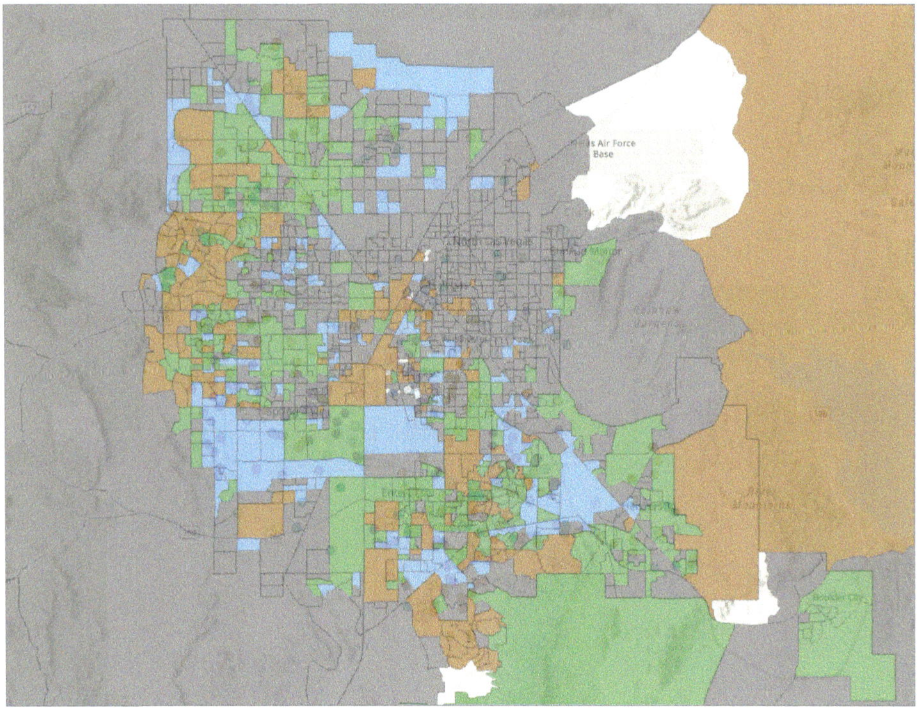

You will explore and turn each of the target groups on or off in the map by using the symbol classes in the Symbology pane.

15. In the **Contents** pane, right-click **Target Group Layer** and click **Symbology**.

16. In the **Symbology** pane, under the **Classes** section, use the check boxes to turn individual **Target Groups** on or off.

    The example shows just the Core and Developmental target groups.

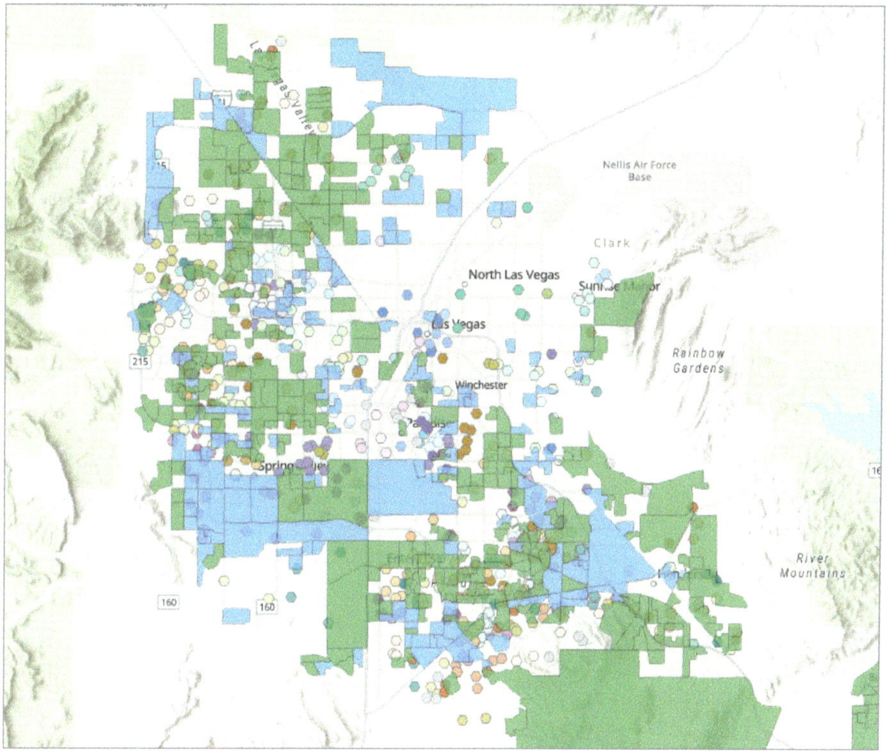


# Tutorial 19-3: Review the segmentation and marketing reports

In this tutorial, you will look at the Segmentation Profile and the Developing Marketing Strategies reports. The reports help you understand who your customers are and which media channels may be best suited for reaching them. This helps you build better marketing campaigns across different media types.

## Examine the segmentation and developing marketing strategies reports

1. Navigate to the **Catalog** pane and expand **Folders** > **Ch 19 - Target marketing**.

2. Hover over the **Tapestry Profile Infographic.pdf** and then click the Path link in the pop-up to open the report in your PDF file viewer, such as Adobe Acrobat or PDF Editor.

esri THE SCIENCE OF WHERE

**Customer Tapestry Profile**

**Top 10 Segments**

TAPESTRY SEGMENTATION esri.com/tapestry

Esri **Tapestry Segmentation** provides a detailed description of America's neighborhoods – U.S. residential areas are divided into distinct segments based on their socioeconomic and demographic composition. Neighborhoods with the most similar characteristics are grouped together, and neighborhoods showing divergent characteristics are separated. Learn more.

### Top 10 Segments

| Tapestry Description | Target Count | Target Percent | Base Count | Base Percent | Penetration | Index |
|---|---|---|---|---|---|---|
| D5 Laptops and Lattes | 3 | 0.50% | 414 | 0.00% | 0.72% | 1052 |
| K3 Loyal Locals | 2 | 0.30% | 558 | 0.10% | 0.36% | 520 |
| H4 Urban Chic | 18 | 3.00% | 6306 | 0.70% | 0.29% | 414 |
| H3 Neighborhood Spirit | 13 | 2.10% | 5281 | 0.60% | 0.25% | 357 |
| D4 Metro Renters | 6 | 1.00% | 3141 | 0.40% | 0.19% | 277 |
| L3 Top Tier | 20 | 3.30% | 12039 | 1.40% | 0.17% | 241 |
| K6 City Greens | 16 | 2.60% | 10119 | 1.10% | 0.16% | 229 |
| D1 Emerging Hub | 27 | 4.40% | 17506 | 2.00% | 0.15% | 224 |
| K8 Burbs and Beyond | 29 | 4.80% | 20586 | 2.30% | 0.14% | 204 |
| K2 Middle Ground | 2 | 0.30% | 1505 | 0.20% | 0.13% | 193 |

Learn about this table

### Segments by Index

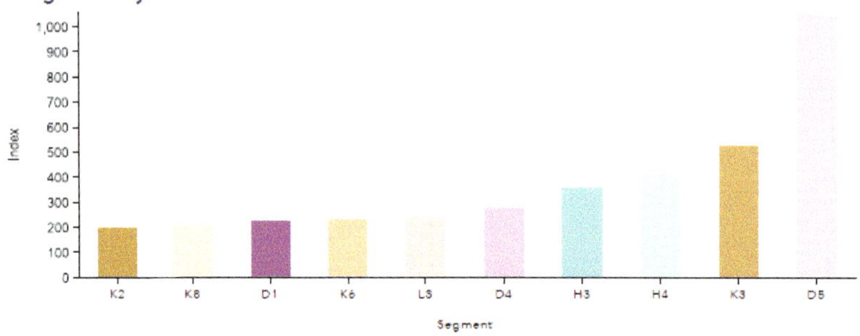


Source: Esri forecasts for 2025.

Page 1 of 7 ©2025 Esri

The Customer Tapestry Profile report has a summary of the top 10 segments within your market. It also has statistics and charts that show how well your dealership is known and how well it is ranked against the market.

Pages 2 and 3 break down these indexes by all segments and LifeModes. For the Las Vegas market, the highest-indexing LifeModes are K: Suburban Shine and L: Premier Estates, which are the two wealthiest LifeModes in Tapestry. Both tend to be older, with larger homes, notably in the suburbs, and high incomes relative to the United States.

LifeMode J: Mature Reflections also indexes high, another older group but less wealthy than Premier Estates and Suburban Shine. This suggests that many of the customers attracted to the dealership may be attracted to higher-end, more expensive vehicles. As they are older, they may also have other family members of driving age for whom they buy cars.

Mature Reflections may want to purchase less expensive cars as they manage retirement funds or plan retirement.

3. Return to ArcGIS Pro. In the **Catalog** pane, hover over **Developing Market Strategies Infographic.pdf** and click the **Path** link in the pop-up to open the report in your PDF file viewer.

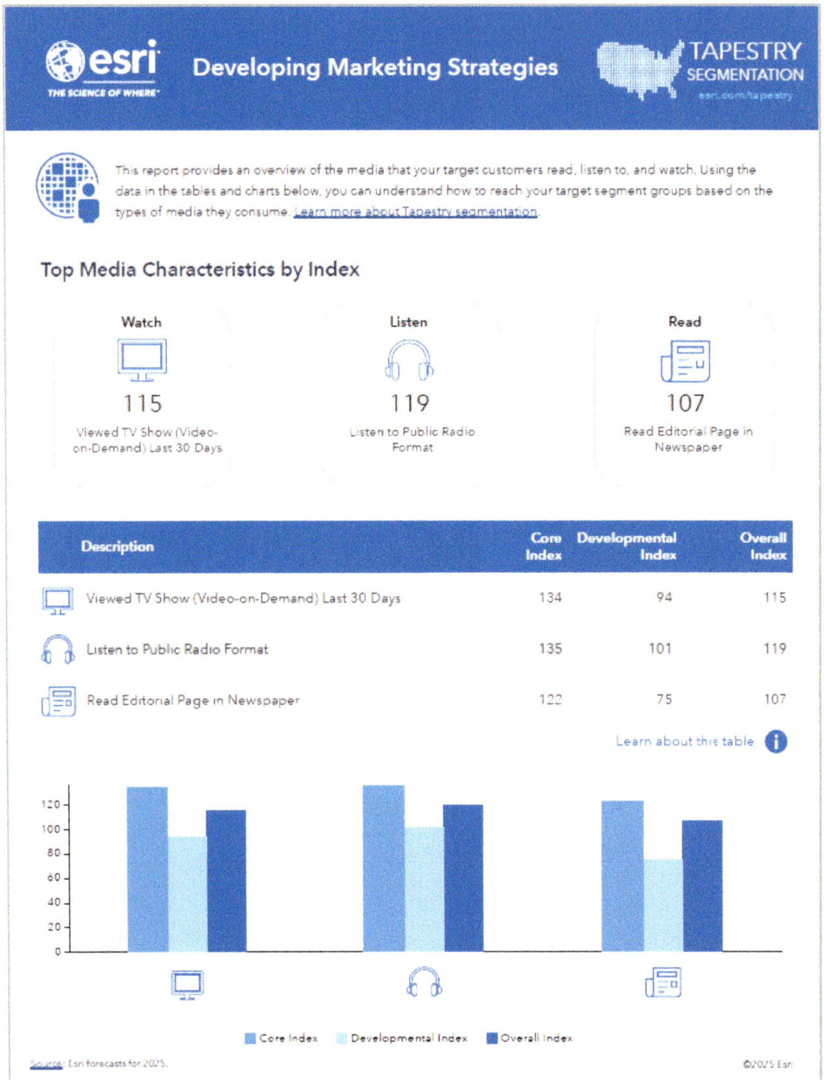


The report shows how Core and Developmental target groups use different media types, including online, TV, radio, and print ads. For the Las Vegas market, the Core target group is much more engaged with all forms of media, with on average almost 30 percent more engagement that the US average, which represents a score of 100.

In contrast, the Developmental target group is much less engaged, being at or below the national average and is much less likely to buy print ads. They only scored 75 on this media type. The report says that both groups will listen to public radio and TV shows and streaming services (video on demand). They will mostly listen to radio ads.

# Summary

In this chapter, you used Business Analyst Pro to model and profile customers using Tapestry to help create better marketing campaigns. You have built maps and reports that can inform your decision-making and maximize the return on investment from your marketing dollars.

## Workflow

1. Load and open an example project.
2. Create target marketing profiles for your customers and the market in which you will execute the marketing campaign.
3. Perform a four-quadrant analysis to filter and refine customer segments.
4. Present the results using maps and reports.
5. Configure the map symbology to maximize understanding and analytical insights.

# CHAPTER 20

# Perform gap analysis with customer data and the Target Marketing Wizard

## Objectives

- Use the ArcGIS Tapestry segmentation system to look at outdoor enthusiasts and sports fans in Indiana.
- Use Four Quadrant Analysis to reclassify the Core and Developmental target groups that are important to make effective marketing plans.
- Quantify market penetration and gaps in the customer base to target campaigns.

## Introduction

Chapter 19 introduced the Target Marketing Wizard, a step-by-step process for analyzing customers and marketing campaign insights.

This chapter combines the wizard with the underlying tools to maximize the flexibility and power of your analysis. Some steps in the wizard will be skipped and replaced with geoprocessing tools, providing a deeper understanding of how you can apply these components and concepts to your own work.

# Tutorial 20-1: Use the Target Marketing Wizard to create customer and market profiles

In this case, you are a marketing expert at a sports and outdoor activity store. You are looking at the Indiana market. You have a sample of customer purchases in the last 30 days. These purchases have been geocoded and assigned a Tapestry segment.

You will use parts of the Target Marketing Wizard to create customer and market profiles and develop a focused Four Quadrant Analysis chart and infographic reports. With these profiles, you will perform penetration and gap analysis to help you target your marketing expenditures and find markets that are not performing well.

> *Important: Target Marketing tools are supported only with locally stored ArcGIS Business Analyst datasets. If you do not have a local dataset, skip this tutorial and go to tutorial 20-2.*

## Download and open the project

1. In your browser, navigate to link.esri.com/Top20BA/Data. Download the **Chapter 20 - Gap analysis** item.
2. Extract the **Ch 20 - Gap analysis** folder to **C:\Top20BusinessAnalyst**.

   **Tip:** This book uses **C:\Top20BusinessAnalyst** for the folder location, but you can use any folder.

3. In the **Ch 20 - Gap analysis** folder, double-click **Gap Analysis.aprx** to open the project in ArcGIS Pro.

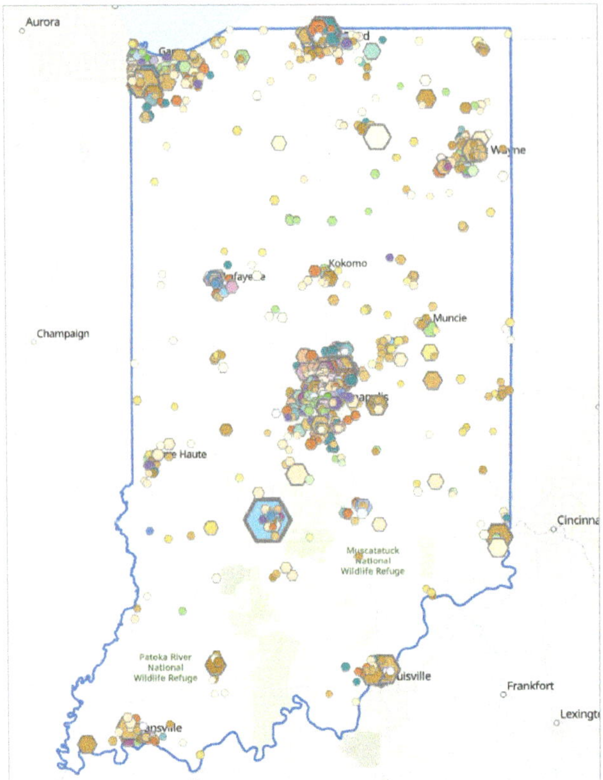


The project opens to an outline of the state of Indiana. The data displays all spending at your stores in the last 30 days. It has been used to make a map that shows the size of the symbol according to how much money was spent and the color of the unique Tapestry segment.

## Create a customer profile for Indiana outdoor enthusiasts

The analysis starts by creating a customer profile based on the 1,371 customers in Indiana. In this chapter, you will use spending data to understand patterns and consumer behavior. Spending behavior is not consistent across markets and varies by Tapestry segment. With better knowledge of how these patterns vary, you can create more focused and successful campaigns.

4. On the ribbon, click the **Analysis** tab. In the **Workflows** group, click the **Business Analysis** gallery and click **Target Marketing Wizard**.

   The Target Marketing Wizard appears.

5. For **Step One: Customer Profile**, apply the following settings:
   - For **Customer Layer**, select **Indiana_Customers**.
   - For **Customer Segmentation Profile**, select **Create new profile**. Type Indiana_Customers_Target_Profile.
   - For **Segmentation Base**, select **Total Households**.
   - For **Volumetric Field**, select **SPEND**.
6. Click **Next**.
7. For **Step Two: Market Area Profile**, apply the following settings:
   - For **Market Area**, select **Area of interest**.
   - For **Area of interest**, next to the search bar, click the map button and select **Indiana**.
   - For **Market Area Segmentation Profile**, make sure **Create new profile** is selected.
   - For **Create new profile**, type Indiana_Base_Profile.
8. Click **Next**.

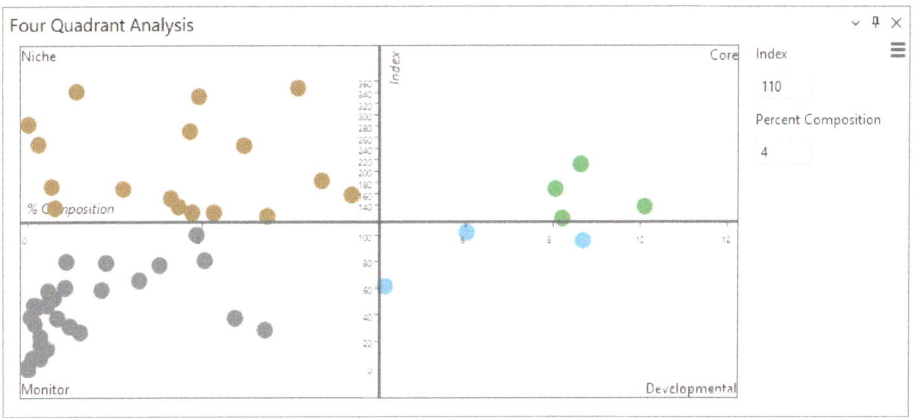


You now have both customer and market profiles, which are combined to create the Four Quadrant chart.

## Configure the Four Quadrant chart to add segments to the Core and Developmental target groups

In this section, you will adjust the axes so that segments from Niche and Monitor are moved to the Core and Developmental target segments. This is being done to increase the number of target segments and better reflect the objectives of the marketing campaign.

The Four Quadrant chart uses a default normalized index of 110 and a 4 percent composition value to divide up the quadrants.

In the Indiana Four Quadrant chart, most segments are less than 2 percent. The penetration index also has a large range, with three segments indexing around 350. This suggests that the split between Niche/Monitor and Core/Developmental is too low at 110.

9. In the **Four Quadrant Analysis** chart, click in the **Index** box and type 120.
10. Click in the **Percent Composition** box and type 2.
11. Click in the **Index** box again to apply both new values and update the chart.

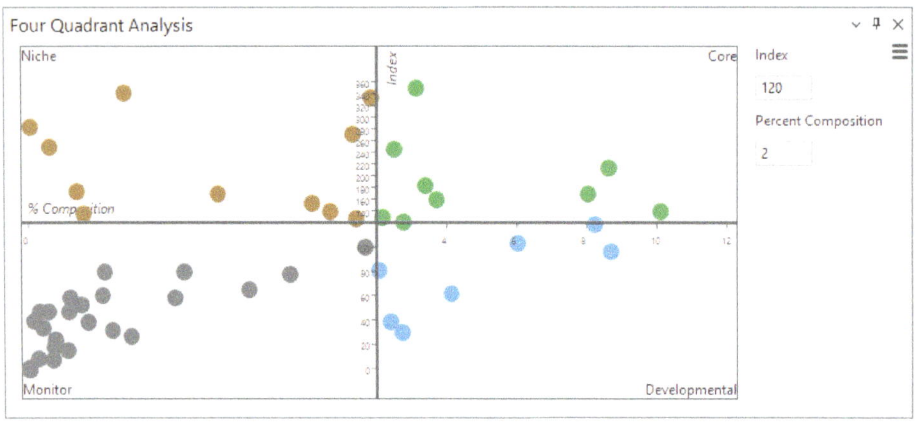


12. In the **Target Marketing** pane, click **Next**.

    You now have a Four Quadrant chart with nine segments in Core and seven in Developmental, totaling 78 percent of the customer base.

## Skip the map generation step

You will skip the map generation part in Step Four: Mapping. These maps can be useful in understanding the analysis but are not needed for this use case.

13. For **Step Four: Mapping**, apply the following settings:
    - For **Geography Level**, select **US.ZIP5**.
    - For **Map Layers**, uncheck all the boxes.

14. Click **Next**.

### Create a volumetric segmentation profile, market area, and gap report

15. For the **Step Five: Reporting** pane, under **Target Marketing Reports**, apply the following settings:
    - Check the **Segmentation Profile** box.
    - Make sure the **Developing Marketing Strategies** box is unchecked.
    - Check the **Market Potential** box.
    - Check the **Market Area and Gap Analysis** box.
    - In the **Export Format** area, make sure the **PDF** box is checked.
16. Click **Next**.
17. For the **Step Six: Summary** pane, review the steps and click **Finish**.

    The wizard has not created any map feature layers, but it has generated the essential customer and market profiles. You also possess a set of informative reports that you will now use to refine and better target customers in support of your marketing campaign.

## Tutorial 20-2: Review the various volumetric, market, and gap reports

In this tutorial, you will look through the reports that were created by using the wizard. Your project will contain two profiles:

- Indiana_Base_Profile.sgprofile
- Indiana_Customers_Target_Profile.sgprofile

And three PDF-format reports:

- Gap Analysis Infographic.pdf
- Market Potential with Volume Infographic.pdf
- Tapestry Profile with Volume Infographic.pdf

## For users who don't have a local dataset

If you do not have a local dataset, follow these steps to download the project:

1. In your browser, navigate to link.esri.com/Top20BA/Data. Download the **Chapter 20 - Gap analysis (Segmented)** item.
2. Extract the **Ch 20 - Gap analysis** folder to **C:\Top20BusinessAnalyst**.
3. In the **Ch 20 - Gap analysis** folder, double-click **Las Vegas Target Marketing.aprx** to open the project in ArcGIS Pro.

## Examine the reports

You will open each of the infographics to examine the top 10 segments and spending patterns for each of the segments and LifeMode groups.

1. In File Explorer, open the **Ch 20 - Gap analysis** folder.
2. Double-click the **Tapestry Profile with Volume Infographic** PDF.

esri THE SCIENCE OF WHERE

Customer Tapestry Profile with Volume

**Top 10 Segments**

TAPESTRY SEGMENTATION
esri.com/tapestry

Esri **Tapestry Segmentation** provides a detailed description of America's neighborhoods – U.S. residential areas are divided into distinct segments based on their socioeconomic and demographic composition. Neighborhoods with the most similar characteristics are grouped together, and neighborhoods showing divergent characteristics are separated. Learn more.

**Top 10 Segments**

| Tapestry Description | Target Count | Target Percent | Total Volume | Average Volume | Average Volume Index |
|---|---|---|---|---|---|
| G1 Shared Roots | 2 | 0.1 | 3,733 | 1866.5 | 210.00 |
| I6 Southern Satellites | 13 | 1.0 | 23,400 | 1800.0 | 202.00 |
| G2 Up and Coming Families | 13 | 1.0 | 20,940 | 1610.8 | 181.00 |
| J2 The Elders | 1 | 0.1 | 1,399 | 1398.7 | 157.00 |
| H4 Urban Chic | 5 | 4.2 | 6,710 | 1342.1 | 151.00 |
| H1 Flourishing Families | 58 | 1.8 | 72,708 | 1253.6 | 141.00 |
| B2 College Towns | 25 | 1.8 | 29,878 | 1195.1 | 134.00 |
| D5 Laptops and Lattes | 3 | 0.2 | 3,296 | 1098.8 | 124.00 |
| K2 Middle Ground | 112 | 8.2 | 122,315 | 1092.1 | 123.00 |
| K3 Loyal Locals | 91 | 6.7 | 97,482 | 1071.2 | 120.00 |

**Segments by Average Volume Index**

Learn about this table

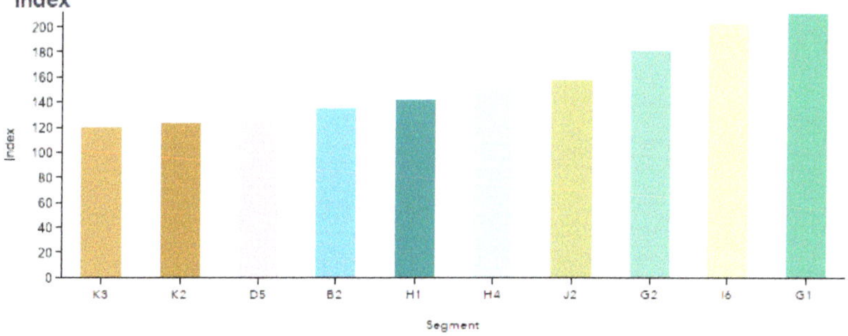


Page 1 of 7 ©2025 Esri

The volume-based Tapestry infographic summarizes the top 10 segments within the market, together with penetration and sales figures for each segment. For the Indiana market, you have just looked at the top three segments: G1 Shared Roots, I6 Southern Satellites, and G2 Up and Coming Families. These three segments have more than 180 customers and spend about $48,000. Although they have a high index, there are only 28 total customers from these three segments in the customer database. On average, these customers have spent $1,717 at the stores in the last 30 days, making them high spenders and desirable customers. G1 Shared

Roots customers are the highest spenders, with an average of $1,866.50 over the same period, as seen in the Average Volume column of the top 10 chart.

The K2 Middle Ground Tapestry segment is the largest of the top 10 segments with 112 customers, who spent $1,092.10 on average.

Immediately, you can see the need to differentiate the campaign into at least two strategies:

- Grow customers in all the top 10 segments
- Make sure that the K2 Middle Ground and K3 Loyal Locals segments remain engaged and continue to spend at the stores.

Overall, LifeModes G: Family Fabric, H: Family Prosperity, and K: Suburban Shine index highest in the top 10 segments. Suburban Shine segments compose 14.9 percent of the top 10 segments, made up of Middle Ground and Loyal Locals segments.

The third page summarizes the segments into LifeMode groups. Books and Boots (LifeMode 2/B) indexes at almost 350, with Premier Estates and Tech Trailblazers making up the top three LifeModes. A detailed look at the behaviors of the Tapestry segments within these LifeModes shows that many of them are outdoors people, are hunters, play sports often, and are active.

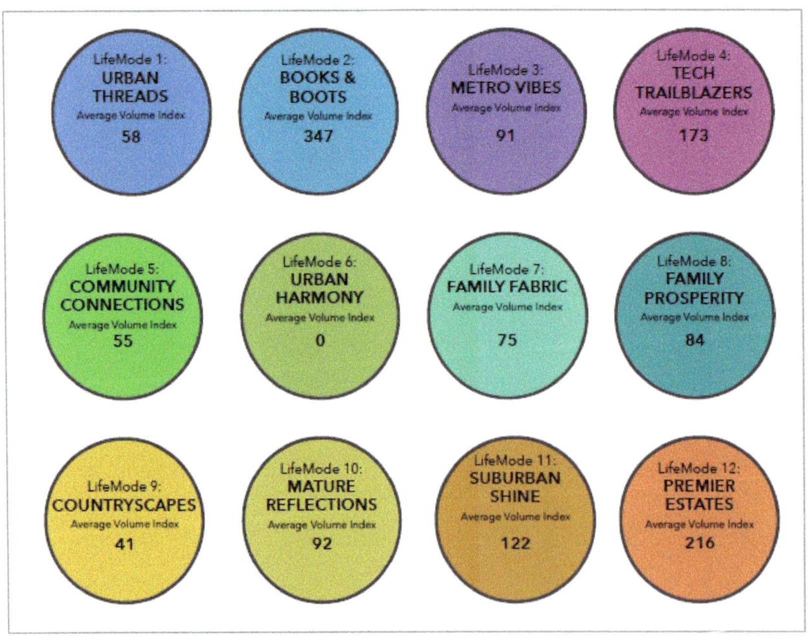

3. In File Explorer, open the **Market Potential with Volume Infographic** PDF.

   The report identifies which markets have the most potential for your marketing campaign based on the likely demand for your product or service. The potential market in Indiana is approximately 3.2 million households, with expected sales of $1.39 million in a similar 30-day period.

   The infographic shows the top five zip codes by index and expected number of customers. It also shows the number of new customers expected based on using zip codes as the geographic polygon type in your wizard-based workflow.

esri THE SCIENCE OF WHERE

**Market Potential Volume Report**

TAPESTRY SEGMENTATION esri.com/tapestry

The Market Potential Volume Report measures the likely demand for a product or service for your market area at a specific geography level. You can use this report to make decisions about where to offer products and services. Learn more.

**Summary**

| | | | |
|---|---|---|---|
| 3,192,183 | 1,391,055.65 | 736.90 | 83.00 |
| Overall Segmentation Base Total | Total Expected Volume | Overall Expected Average Volume | Overall Expected Volume Index |

**Top Areas by Expected Volume** Learn about this table

| Geography ID | Geography Name | Segmentation Base Total | Expected Customers | Expected Volume | Expected Average Volume | Expected Volume Index |
|---|---|---|---|---|---|---|
| 47906 | West Lafayette | 27,457 | 38 | 26,616.91 | 700.44 | 79.00 |
| 46032 | Carmel | 22,061 | 25 | 19,237.11 | 769.48 | 87.00 |
| 47401 | Bloomington | 18,695 | 20 | 16,085.87 | 804.29 | 90.00 |
| 46033 | Carmel | 14,536 | 19 | 15,004.82 | 789.73 | 89.00 |
| 46307 | Crown Point | 27,338 | 16 | 14,835.09 | 927.19 | 104.00 |

**Top Areas by Expected Volume Index** Learn about this table

| Geography ID | Geography Name | Segmentation Base Total | Expected Customers | Expected Volume | Expected Average Volume | Expected Volume Index |
|---|---|---|---|---|---|---|
| 46239 | Indianapolis | 12,809 | 6 | 6,891.83 | 1148.64 | 129.00 |
| 46113 | Camby | 6,238 | 3 | 3,291.36 | 1097.12 | 123.00 |
| 47725 | Evansville | 7,404 | 4 | 4,272.70 | 1068.18 | 120.00 |
| 46231 | Indianapolis | 3,963 | 2 | 2,113.08 | 1056.54 | 119.00 |
| 46818 | Fort Wayne | 9,341 | 4 | 4,151.80 | 1037.95 | 117.00 |

Source: Esri forecasts for 2025.

You can explore more recommendations and learn about the results in the Market Potential with Volume Infographic by clicking the hyperlinks.

4. In File Explorer, open the **Gap Analysis Infographic** PDF.

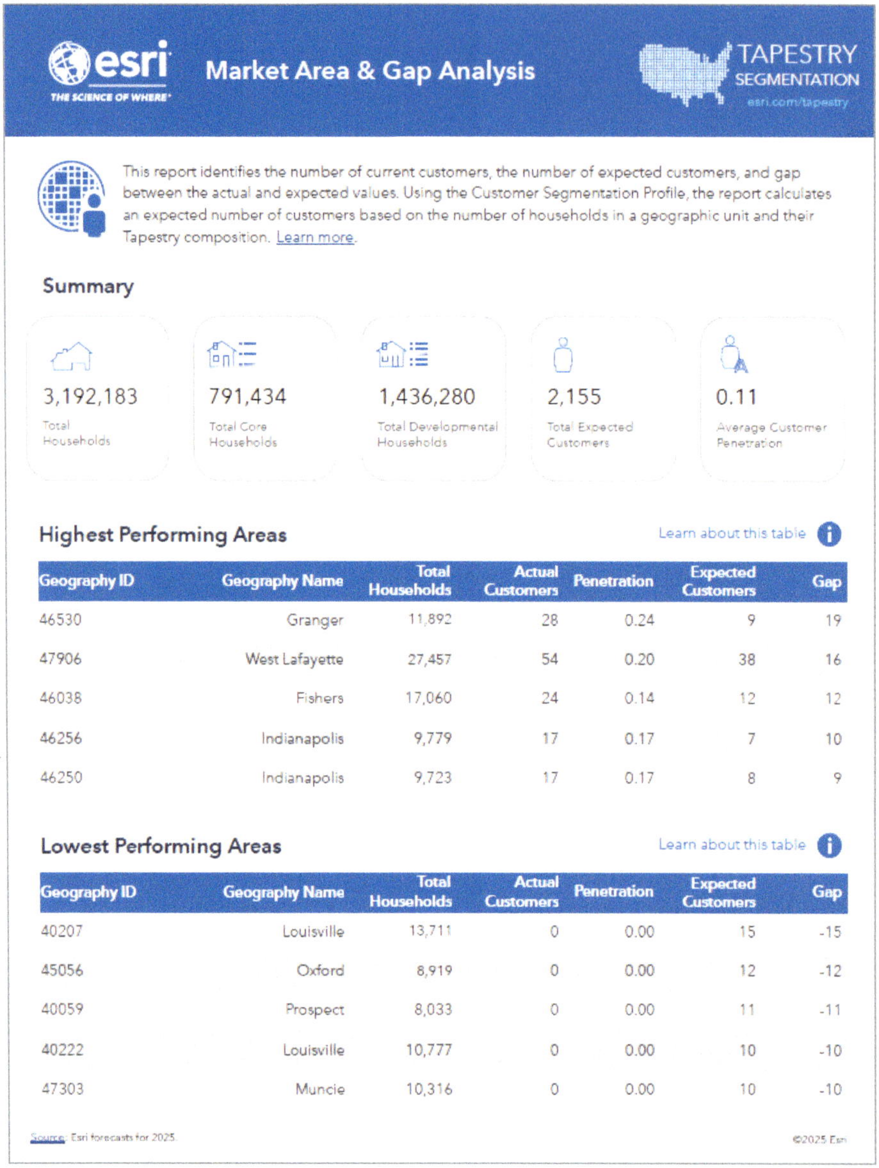

esri THE SCIENCE OF WHERE

**Market Area & Gap Analysis**

TAPESTRY SEGMENTATION esri.com/tapestry

This report identifies the number of current customers, the number of expected customers, and gap between the actual and expected values. Using the Customer Segmentation Profile, the report calculates an expected number of customers based on the number of households in a geographic unit and their Tapestry composition. Learn more.

**Summary**

| 3,192,183 | 791,434 | 1,436,280 | 2,155 | 0.11 |
|---|---|---|---|---|
| Total Households | Total Core Households | Total Developmental Households | Total Expected Customers | Average Customer Penetration |

**Highest Performing Areas** — Learn about this table

| Geography ID | Geography Name | Total Households | Actual Customers | Penetration | Expected Customers | Gap |
|---|---|---|---|---|---|---|
| 46530 | Granger | 11,892 | 28 | 0.24 | 9 | 19 |
| 47906 | West Lafayette | 27,457 | 54 | 0.20 | 38 | 16 |
| 46038 | Fishers | 17,060 | 24 | 0.14 | 12 | 12 |
| 46256 | Indianapolis | 9,779 | 17 | 0.17 | 7 | 10 |
| 46250 | Indianapolis | 9,723 | 17 | 0.17 | 8 | 9 |

**Lowest Performing Areas** — Learn about this table

| Geography ID | Geography Name | Total Households | Actual Customers | Penetration | Expected Customers | Gap |
|---|---|---|---|---|---|---|
| 40207 | Louisville | 13,711 | 0 | 0.00 | 15 | -15 |
| 45056 | Oxford | 8,919 | 0 | 0.00 | 12 | -12 |
| 40059 | Prospect | 8,033 | 0 | 0.00 | 11 | -11 |
| 40222 | Louisville | 10,777 | 0 | 0.00 | 10 | -10 |
| 47303 | Muncie | 10,316 | 0 | 0.00 | 10 | -10 |

Source: Esri forecasts for 2025. ©2025 Esri

This report shows how many customers there are and how many customers you expect to have in each area. The gap between the actual and expected customer values is calculated using the Customer Segmentation Profile. The expected number of customers is based on the number of households in a geographic unit and their Tapestry composition, whereas the actual number of customers is based on summarizing all the customers within the geographic unit, which for your analysis was zip codes.

Reading the report, you can see that there are 3,192,183 total households in the Indiana market, of which 791,434 households are in the Core target group. This is just under 25% or one-quarter of all households in Indiana. A total of 1,436,280 households are in the Developmental target group, representing 45% of all households in Indiana. Together, your main customer target groups cover 70% of Indiana's households.

The Gap Analysis Infographic also shows the best- and worst-performing zip codes. It shows where you should have more customers based on the penetration and Tapestry segment characteristics and where you are doing better than expected with more customers than expected.

The Gap Analysis Infographic says you should have 2,155 customers from Indiana in the last 30 days. This is a difference of 784 customers, which is the difference between the number of customers predicted by the gap analysis using all Tapestry segments in the profile for Indiana and the number of customers in the Indiana sample data. The gap of 784 would be one of the metrics for your marketing campaign as this is the minimum number of new customers your campaign should generate.

# Tutorial 20-3: Map customer and market gaps

In this tutorial, you will use tools to perform a market area gap analysis. With the results from the analysis, you will filter and symbolize them based on zip codes.

## Use the Analyze Market Area Gap geoprocessing tool

In this section, you will use the Analyze Market Area Gap tool to create a new layer of market conditions and customer statistics.

1. Return to ArcGIS Pro.
2. On the ribbon, on the **Analysis** tab, in the **Geoprocessing** group, click **Tools**.
3. In the **Geoprocessing** pane, click the **Toolboxes** tab. Expand **Business Analyst Tools** > **Target Marketing** and click **Analyze Market Area Gap**.

4. In the **Analyze Market Area Gap** tool, apply the following settings:

   - For **Customer Layer**, select **Indiana_Customers** .
   - For **Target Profile**, click the browse folder icon. In the **Ch 20 - Gap analysis** folder, click **Indiana_Customers_Target_Profile.sgprofile**. Click **OK**.
   - For **Base Profile**, browse to the **Ch 20 - Gap analysis** folder and select **Indiana_Base_Profile.sgprofile**. Click **OK**.
   - For **Geography Level**, select **ZIP Codes (US.ZIP5)**.
   - For **Output Feature Class**, type Indiana_Market_Area_Gap.
   - For **Target Group**, browse to the **Ch 20 - Gap analysis** folder and select **New Target Group.sgtargetgroup**. Click **OK**.
   - For **Core Target**, select **Core**.
   - For **Developmental Target**, select **Developmental**.
   - For **Boundary Layer**, select **Indiana**.

5. Click **Run**.

6. In the **Contents** pane, turn off the **Indiana_Customers** layer.

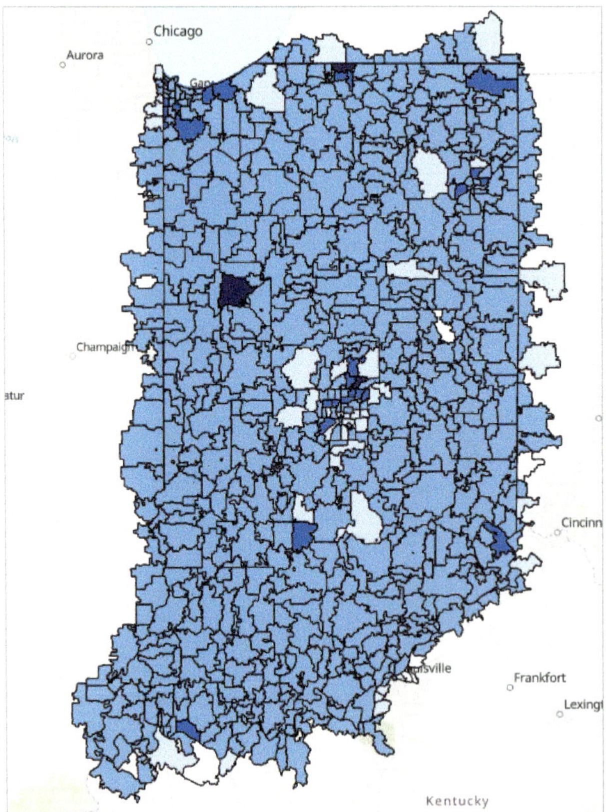

You now have a feature layer symbolized by the size of the market area gap. The map shows a market area gap of between −9 and 19 customers across the zip codes in the Indiana market.

## Use a definition query to remove zip codes that are not in Indiana

By choosing Indiana as the target area, you also created a map with zip codes that touch Indiana. This is because when using the Target Marketing tools, ArcGIS Business Analyst assumes that you want to include geographies near or touching your market area.

For this analysis, you will use a definition query to remove all zip codes that are not in Indiana.

7. In the **Contents** pane, right-click the **Indiana_Market_Area_Gap** layer and click **Properties**.
8. In the **Layer Properties** window, click the **Definition Query** tab.
9. Click **New definition query**.
10. Under **Query 1**, for the **Where** field, scroll to the bottom of the list of variables and select **RG_NAME**.

    The criteria operator automatically fills in as is equal to.
11. In the next empty cell, select **Indiana**.

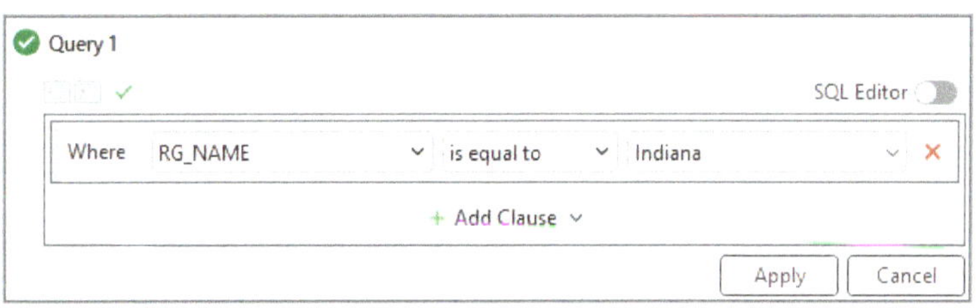


12. Click **OK**.

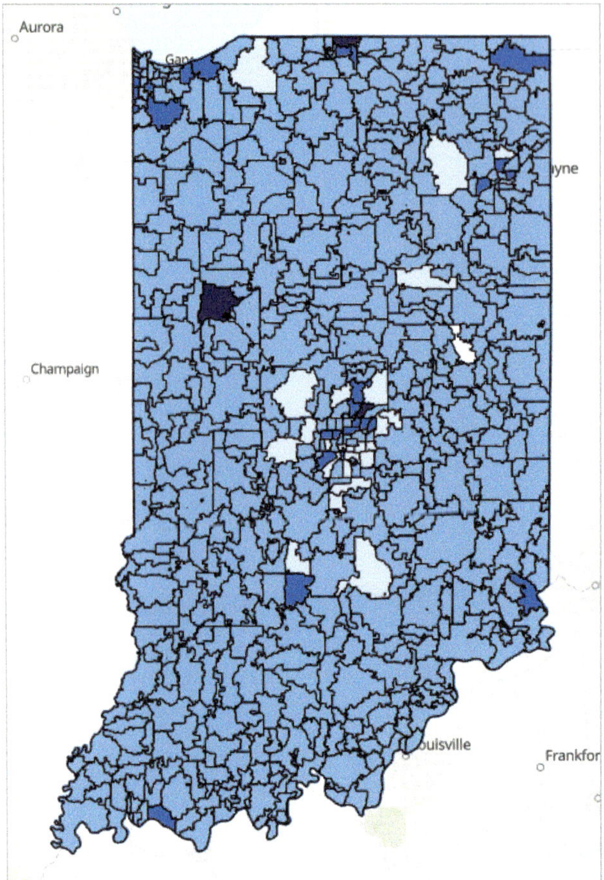


You now have focused the Market Area Gap Analysis feature layer solely on Indiana zip codes.

## Change the symbology of the gap analysis map

13. In the **Contents** pane, right-click the **Indiana_Market_Area_Gap** layer and click **Symbology**.

14. In the **Symbology** pane, apply the following settings:

    - For **Method**, select **Equal Interval**.
    - For **Classes**, select **10**.
    - For **Color scheme**, open the list and select **Choose a color scheme**. In the **Choose a color scheme** window, type red yellow blue. Scroll down and select **Red-Yellow-Blue (Continuous)**. Click **OK**.

15. On the ribbon, click the **Feature Layer** contextual tab. In the **Effects** group, change the **Transparency** to 30.0%.

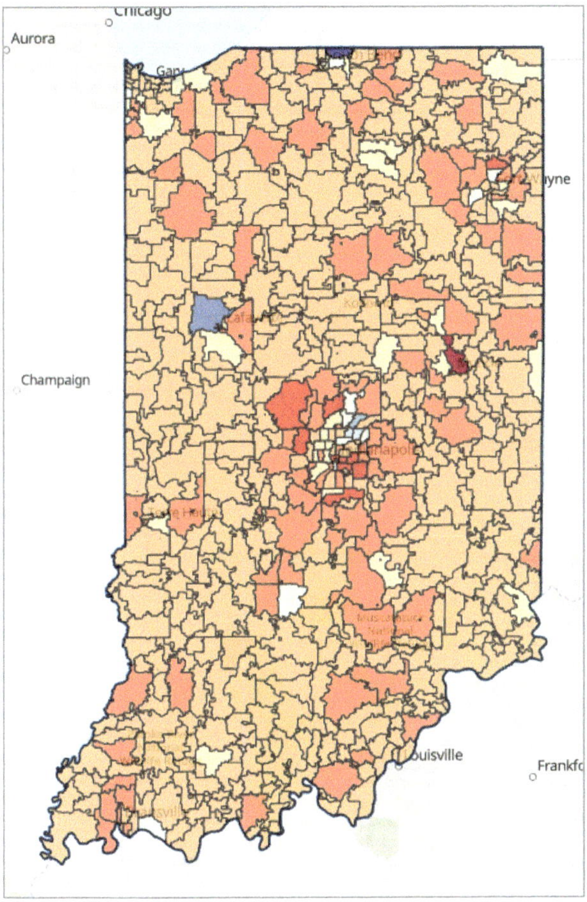


You now have a zip code map symbolized from red to blue, where red zip codes have a negative gap, less customers than expected and blue zip codes have more customers than expected.

This map lets you explore the many patterns that exist and builds on the information in the Market Area Gap Analysis Infographic. Understanding and quantifying the patterns the map reveals will guide you in shaping your marketing strategy. Following are two example areas where zip codes show diverging patterns:

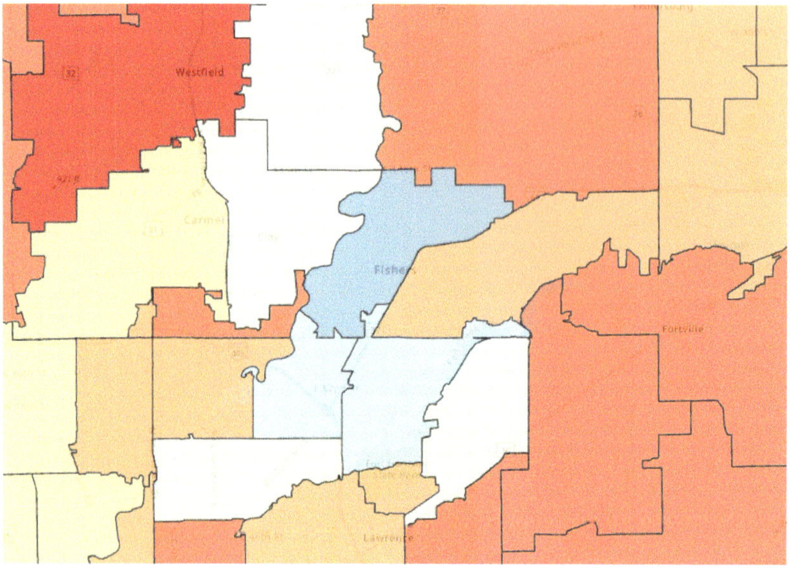


In Northern Indianapolis, neighborhoods around Fishers have more customers than expected compared with those near Westfield.

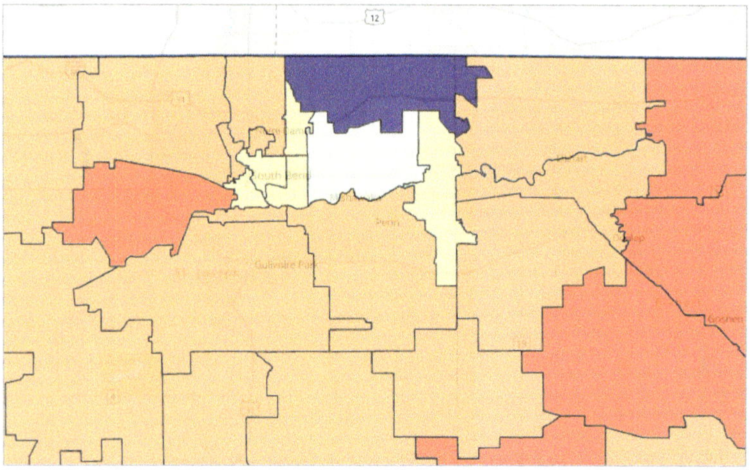


At the top of the state, the Granger zip code 46530 has three times more customers than expected. The nearby Plymouth zip code 46563 has no customers, similar to many of the zip codes around South Bend, which is home to Notre Dame University.

Feel free to explore any of the attributes generated by the Analyze Market Area Gap geoprocessing tool, to understand penetration, actual and expected customers, or the distribution of each Tapestry segment by zip code.

## Summary

In this chapter, you used Business Analyst Pro to profile customers and identify market gaps using Tapestry to help create better marketing campaigns. You used information from reports, analysis, and maps to help you make decisions and get the most of your marketing dollars.

### Workflow

1. Load and open an example project.
2. Create target marketing profiles for your customers and the market in which you will execute the marketing campaign.
3. Customize the Four Quadrant chart to meet the needs of the campaign.
4. Identify market gaps and differences between actual and expected customers.
5. Present the results using maps and reports.
6. Configure the map symbology to maximize understanding and analytic insight.

# DATA SOURCE CREDITS

## Chapter 17

Courtesy of Esri ArcGIS Business Analyst team.

## Chapter 18

Courtesy of Esri ArcGIS Business Analyst team.

## Chapter 19

CUSTINFO database. Courtesy of Esri and Helen Thompson.

## Chapter 20

CUSTINFO database. Courtesy of Esri and Helen Thompson.

# ABOUT ESRI PRESS

Esri Press is an American book publisher and part of Esri, the global leader in geographic information system (GIS) software, location intelligence, and mapping. Since 1969, Esri has supported customers with geographic science and geospatial analytics, what we call The Science of Where®. We take a geographic approach to problem-solving, brought to life by modern GIS technology, and are committed to using science and technology to build a sustainable world.

At Esri Press, our mission is to inform, inspire, and teach professionals, students, educators, and the public about GIS by developing print and digital publications. Our goal is to increase the adoption of ArcGIS and to support the vision and brand of Esri. We strive to be the leader in publishing great GIS books, and we are dedicated to improving the work and lives of our global community of users, authors, and colleagues.


## Acquisitions

Stacy Krieg
Alycia Tornetta
Jenefer Shute
Katie Gezi

## Product Engineering

Craig Carpenter
Maryam Mafuri

## Editorial

Carolyn Schatz
Mark Henry
David Oberman

## Production

Monica McGregor
Victoria Roberts

## Sales & Marketing

Eric Kettunen
Sasha Gallardo
Beth Bauler

## Contributors

Christian Harder
Matt Artz

## Business

Catherine Ortiz
Jon Carter
Jason Childs


**Was this book valuable and helpful?**

If so, please consider leaving an honest review on Amazon, Goodreads, other platforms, and in the Esri Community (community.esri.com). Customer reviews are a great way to share books you found valuable with your community, and they help us continue to publish the world's best GIS books!

# Related titles

**The Power of Where**
Jack Dangermond, Esri, and The GIS Community
9781589486065

**Getting to Know ArcGIS Enterprise**
Jon Emch, Diana Muresan, and Travis Ormsby
9781589487932

**Spatial Business**
Competing and Leading with Location Analytics
Thomas A. Horan, PhD, James B. Pick, PhD, and Avijit Sarkar, PhD
9781589485334

**Top 20 Essential Skills for ArcGIS Online**
Craig Carpenter, Jian Lange, and Bern Szukalski
9781589487802